THE GLOBAL
ENVIRONMENTAL CRISIS

The Global Environmental Crisis presents a new perspective on our inattention and inaction in the face of a major crisis.

We cannot proceed without scientific knowledge, but we cannot exclusively rely on it. What we need, in addition to scientific knowledge, is utopian imagination to make us understand the nature of the crisis and to suggest an alternative vision of a viable future.

This book is an essential resource for students and instructors across the social sciences, especially sociology and environmental studies. It will also be a crucial and accessible text for general readers interested in climate change and how to imagine a better world for themselves and future generations.

John Lie is Distinguished Professor of Sociology at the University of California, Berkeley, where he teaches social theory. His books include *Modern Peoplehood* (2004) and *Japan, the Sustainable Society* (2021).

THE GLOBAL ENVIRONMENTAL CRISIS

The Limitations of Scientific Knowledge and the Necessity of Utopian Imagination

John Lie

Routledge
Taylor & Francis Group

NEW YORK AND LONDON

Designed cover image: Earthrise, NASA/Bill Anders

First published 2025
by Routledge
605 Third Avenue, New York, NY 10158

and by Routledge
4 Park Square, Milton Park, Abingdon, Oxon, OX14 4RN

Routledge is an imprint of the Taylor & Francis Group, an informa business

© 2025 John Lie

ISBN: 978-1-032-88724-1 (hbk)
ISBN: 978-1-032-87396-1 (pbk)
ISBN: 978-1-003-53940-7 (ebk)

DOI: 10.4324/9781003539407

Typeset in Sabon
by Deanta Global Publishing Services, Chennai, India

for Chloé

CONTENTS

ACKNOWLEDGMENTS

Having spent my early childhood in the late 1960s in Tokyo—then at the forefront of pollution-induced disasters and therefore of pollution studies as well—I was alerted early and insistently to the potential hazard of an environmental catastrophe. I participated in the margins of ecological politics and dabbled at environmental studies, crafting an article here and churning a review there. It is in the last decade, however, that I could no longer suppress a sense of urgency. Terrible phenomena that were supposed to happen after my demise began to occur and afflict my life: at the most mundane level, northern Californian wildfires occasioned repeated blackouts. For a creature of comfort, several days without a warm shower concentrates the mind. I did the un-American act of forswearing car ownership. I began to offer (unpopular) courses on the global environmental crisis. Still I could not put off thinking and reading about it. All said and done, the result is this book.

I wish to acknowledge the University of California, Berkeley, my employer for the past two decades. As much as I love to complain about the bloated bureaucracy and systematic dysfunctions of the University, it has provided me with time and other precious resources to complete the book. Thanks especially to the people at the Department of Sociology and the Institute of East Asian Studies. Given the source of my proto-ecological consciousness, it may be fitting that I completed the manuscript in May 2023 in Tokyo, where I benefited from my affiliation with Tokyo College (Institute for Advanced Study, University of Tokyo) in May 2023. I thank Director Masashi Haneda and his staff.

Several people read an earlier draft of this book, and I appreciated the suggestions by Anne E. Gendler, Janna Huang, and Mary Shi. Thanks also to the three anonymous reviewers for Routledge, where Mike Gibson and Saraswathi Prasanthi ably shepherded the manuscript.

I am also grateful to Jinsoo An, Yoko (and Daniel, Hiroshi, and Kyoko) Aoyama, Gillian Edgelow, Kumi Sawada Hadler, Hiroshi Ishida, Perry and Judy Mehrling, Gen (and Kaori) Miyagaki, Yuki (and Humuna) Nakamura, Laura Nelson, Jay (and Victoria and Pearl) Ou, Leslie Salzinger, Sawako Shirahase, Charis Thompson, and (the veritable Santa Claus) Jun Yoo.

This book is dedicated to Chloé, and therefore by extension to Charlotte.

PREFACE

We all know, in the abstract, that we will die, but death happens concretely—indeed, it has happened only—to others. It's a matter for our future, and not present, self. Death cannot cogently be an experience in and of life. But what of an existential threat that has been devastating the world as we know and experience it, such as the global environmental crisis?

Given the prospect of worldwide calamity, an extraterrestrial visitor might expect human beings to take a leaf from Hamlet: "If it be now, 'tis not to come: if it be not to come, it will be now: if it be not now, yet it will come: the readiness is all." Many people suspect that the world is slouching toward an ecological disaster. The plethora of newsfeeds creates a climate at once of inevitability and insouciance; all the facts and figures are as alluring as a stack of UN reports, to be piled one atop another in a dusty digital folder. Without belittling mundane and heroic efforts to alleviate and avoid the ravaging of nature and many who profess concern and dutifully recycle their waste while suffering in turn from eco-anxiety, most people remain inured to the problem and continue to live as they have. Procrastination is the royal road.

This book proposes an answer to the paradox of recognized danger and widespread indifference. Why are we so slow to come to grips with an impending disaster and poorly prepared to do something about it? Why the inattention and inaction?

Three important caveats. First, I dispense with the preliminary task of proving the likelihood of a global environmental catastrophe in this century. There are many reports and books that sound the alarm loud and clear, the most prominent being the periodic assessment reports from the Intergovernmental Panel on Climate Change (IPCC). If only by dint of Lotka's Law—that natural selection will enhance "the energy flux through the system [to] a maximum" (Lotka 1922, 148)—we are clearly in trouble (cf. Smil 2017, 296, 441). The coming—indeed, ongoing—calamity is a hypothesis that I hope against hope that it's wrong, but I won't expatiate on facts and figures, trends and contours (for accessible accounts, see Frankopan 2023, Kolbert 2021, Wallace-Wells 2019; for academic expositions, see Gollier 2019, Pindyck 2022, Smil 1993). *Pari passu* I don't delimit terms such as "crisis" or "catastrophe" beyond the banal sense in which they denote distressing or dreadful states. Suffice it to say that the global environmental crisis means that we cannot sustain our way of life. Surely climate change in and of itself poses a dire threat (for a projection of what each degree of global warming will mean for us, see Lynas 2020.) Unfortunately, global warming is but one facet of the global environmental crisis: think only of the vast outpouring of waste—from spent plutonium rods to unrecycled plastic bottles—and other old-fashioned pollutants (cf. Passmore 1974, Franklin-Wallis 2023). We live amidst a massive die-off of our fellow creatures—the sixth extinction (Kolbert 2014)—while our anthropocentric cast of mind resists our affinity with plants, insects, and even other animals. An adequate conspectus of ecological ills would be long and depressing.

Second, this book does not describe or explain the global environmental crisis. There are many fine accounts that probe the concatenation of forces, major and minor, that are intimately intertwined in its making. Although it is almost always a good idea to probe the nature of the problem thoroughly before coming up with a solution, this is not always the case. Facing an urgent and escalating threat, it may behoove us to pursue multiple solutions, rather than insistently theorizing or analyzing the problem.

Finally, this book is not a comprehensive treatise on the myriad causes that contribute to our inattention and inaction. For a truncated *tour d'horizon* of modes of denial and procrastination, please turn to the appendix.

In this book, I focus on knowledge—specifically, scientific knowledge and scientific worldview—and habitual inaction. Science offers the most prestigious and privileged mode of knowledge, but it is alas limited. Its fecundity depends on an extensive division of intellectual labor, including the chasm between the study of nature and that of society. As a critical endeavor, science progresses by confraction and construction, but it is not always adept at proffering a conspectus. Therefore, the synthetic view that we need emerges belatedly. As a democratic form of knowledge—in which a community of professional investigators assesses truth claims via peer review and other mechanisms—science is perforce slow. It is not surprising, then, that alarms are sounded by journalists, essayists, novelists, and others who depend on, but are not part of, the machinery that generates scientific knowledge. That the work of synthesis arises slowly may be fine for most phenomena, but what if it is an existential crisis whose coming appears to be accelerating? A raven-messenger mired in palaver is not what we need.

The Scientific Revolution—usually dated to the seventeenth century—transformed our understanding of the universe. Natural scientists have explored vast horizons, with great success, but we are left with an outdated worldview. In particular, *certainty* or *determinism* holds sway, rather than the probabilistic or stochastic outlook. Though seemingly arcane, this older epistemology has manifold consequences on our understanding, such as the propensity to regard nature's workings as separate from our roles and activities (which are in turn determined). An environmental catastrophe, in this reckoning, is beyond human control. That is, *inattention* and *indifference* are regnant. Exceptions rely on another outdated mythology: the belief that science and technology will solve the problem. Techno-utopianism, however soothing an ideology, is a misplaced faith, especially as it is precisely the confidence in scientific and technological progress that has generated the environmental crisis in the first place.

Finally, scientific knowledge and our antiquated scientific worldview present a misleading understanding of human action. Far from being rational optimizers who wallow in choices and therefore experience freedom—the product of the scientized social sciences—we are creatures of habit for whom conscious action is rare. Indeed, habitual *inaction* is the norm in human existence. Put differently, we cannot hope to ingest scientific knowledge and expect enlightened individuals to spring instantaneously to (rational) action. Instead, we need other sources of information and inspiration—*utopian imagination*—if we are to avoid the impending disaster. It does not take a leap of faith to embrace and enact a simple ethical injunction, but the sciences do not offer that sort of knowledge or judgment. We also need imagination and wisdom.

Before I proceed, it is also worth bemoaning the Eurocentric—more accurately, US-centric—cast of this book. This may be an instance of bad faith, especially as the brunt of the current crisis is borne by the less privileged people in the Global South. Ecological imperialism proceeds hand in hand with intellectual imperialism (cf. Crosby 1986; Wallerstein 2006, 51–52). All the same, the frontiers of science and technology remain ensconced in rich countries, and my argument therefore focuses on these regions of the world.

Let me reiterate that my presumption may be wrong. Geoengineering or nuclear physics may yet triumph and validate the ruling religion of science and technology. Techno-scientific advances—carbon dioxide-consuming bacteria, portable nuclear fusion, or space colonization—may save us. In this regard, the environmental economist Nordhaus (2021) expatiates on the feasibility of space colonization more than environmental politics or the topics discussed in this book. The stochastic worldview does not offer a predetermined telos of doom and despair, but it is hard to resist the gnawing sensation that we on Spaceship Earth are being piloted and assisted by a crew that seeks to steer full blast ahead against reports about giant obstacles and rough patches on the horizon. The image of the *Titanic*—the modern symbol of technological hubris—is difficult to dislodge. Had we but world enough, and time, but alas we have only one world and not much time.

The enormity of the problem threatens to extinguish hope, and some counsel, as in Dante entering the inferno, to abandon hope. There is now a library of books that assume the imminent end of our way of life (e.g., Servigne and Steven 2015), and some are written as postmortems on where we went wrong (e.g., Oreskes and Conway 2014). As Vollmann's (2018, 13) proleptic lament about our vanishing world summarizes the situation: "In the end I did nothing just the same, and the same went for most everyone I know." Science fiction writers have been there earlier; the dystopian literature on post-apocalyptic life is about as good a looking glass as any (Ballard 1964, Butler 1993, Robinson 2020). Wistfully, then, we are nostalgic for a future that might have been, all the while immobilized by the thought that nothing much can be done.

We are intellectually lazy and politically inactive most of the time, but the future is *not* set: our course is far from being irreversible. There is "the obligatory note of hope" (Offill 2020, 67) that is the basso ostinato of our lived life. It is often assumed that cataclysmic events and processes of world history—the decline and fall of the Roman Empire, for instance—are paradoxically inevitable and invisible: it was fated, and yet no one saw it coming. However, there were contemporaries who worried about it and warned others—and not just Polybius—and the same can be said about global warming and other manifestations of the environmental crisis (cf. Weart 2008). Indeed, already by the 1970s, the basics of climate change were known and efforts were made to rectify the situation by *Republican*-Party politicians in the United States (Rich 2019). Yet, if we regard IPCC's first report in 1990 as the second coming of Cassandra, then we are still left with the disconcerting recognition that more carbon dioxide has been emitted since 1990 than in the entire history of humanity until then (Institute for European Environmental Policy 2020). Reflecting on our knowledge and action is a small, though a necessary, step. Ingrained habits of inattention and inaction can change. We all play our small part in reproducing or reforming the world.

1

BELATED KNOWLEDGE

In the late 1950s, Revelle and Suess (1957, 18) resurrected the nineteenth-century hypothesis that "an increase of carbon dioxide [would] increase the average temperature near the earth's surface." They proposed what we call global warming. As they acknowledged, several scientists had identified the same phenomenon in the nineteenth century (in retrospect Arrhenius's 1896 paper is especially prescient) (Weart 2008, viii). Neither Arrhenius nor his contemporaries who made a similar claim—nor for that matter Revelle and Suess—made a huge splash at the time, but growing ecological consciousness and the environmental movement recognized climate change as a clear and present danger to humanity.

In 1989—nearly a century after Arrhenius's article and more than three decades after Revelle and Suess's—McKibben published a bestselling book, *The End of Nature*, that had been serialized in *The New Yorker*. Based on a generation of research since Revelle and Suess's article, McKibben's book alerted the anglophone reading public to the deleterious consequences of global warming. Stressing that neither time is as "long" nor Earth as "big" as people believe, he made a shocking claim about the end of nature. Although there had been a series of warnings about pollution and environmental decay—think only of the US tradition of nature writing (from Thoreau to Muir, Carson, and beyond) that McKibben (2008, xxii–xxiii) calls "America's single most distinctive contribution to the world's literature"—it was the first widely read account of global warming's potential peril to human survival.

Three decades after *The End of Nature*, Wallace-Wells (2019, 7) opened *The Uninhabitable Earth* by asserting, "It is worse, much worse, than you think." It would be challenging to offer a positive spin on the deteriorating situation he describes: rapidly thawing glaciers,

DOI: 10.4324/9781003539407-1

biodiversity loss, and other manifestations of accelerated global warming. Clearly, argues the author, we are in trouble.

It is possible to narrate the foregoing chain of publications as a tale of triumph. Innovative scientists made a pathbreaking finding that was disseminated, slowly but surely, to the wider public and the policymaking elite. Although three decades separated Revelle and Suess's paper from McKibben's book, the underlying reality was not as ominous in the 1950s. Furthermore, seventy-five years later, there is now a scientific consensus and people are working on solutions to mitigate and possibly solve the constellation of threats (cf. Edwards 2010). Indeed, there may be a magic bullet or two that will shatter the wall of despair elicited by climate change in particular and the environmental crisis in general. There is no reason to read this book (or to write it, for that matter). Science will save us; progress will continue.

There is, however, another way to think about the narrative. Far from denouncing the impending disaster and sounding the alarm, scientists have only slowly—too slowly—issued warnings that were in turn temperate and at times ambiguous.

Consider the series of reports by the Intergovernmental Panel on Climate Change (IPCC) (cf. De Pryck and Hulme 2023). These documents synthesize extant scientific knowledge and are widely hailed as the gold standard—however anachronistic the term—of climate science. Each report was replete with what seemed at the time to be dark forebodings about humanity's future. Yet a more sobering trend is that worst-case scenarios have exacerbated over the past two decades, as was clearly recognized in the Third Assessment Report: "Temperature increases are projected to be greater than those in the Second Assessment Report" (IPCC 2001, 8). What the IPCC had assumed would be a mean world temperature increase of 1°C by 2025 (IPCC 2001) occurred a decade earlier than predicted (IPCC 2021, 142).

A similar phenomenon of scientific underestimation can be seen in other reports (e.g., Stern 2006, Stern 2015, 303). As in Wallace-Wells's portentous opening, the reality was worse, much worse than what scientists had been saying. Popular writings that depend on scientific findings inevitably lag behind, and the more scientifically literate and responsible the writer, the more likely she is to soften her message. The reality may be worse, much worse, than what Wallace-Wells and others had written, however. His dire diagnosis transformed into a "permanent emergency" after the publication of the book, as some of his potential future hazards occurred earlier than had been projected (Wallace-Wells 2021).

An inevitable lag between discovery and action—think of Fleming's accidental discovery of penicillin and the taming of tuberculosis—may be unfortunate but necessary. Who would recklessly dispense with peer reviews and controlled experiments? Who wants to be the first to try an untested vaccine? Carefulness makes scientific knowledge and its application a most impressive achievement. But it may be catastrophic when a phenomenon, such as the deteriorating environment, is not only complex but also *accelerating*. The environmental crisis is complex in the sense that it is about not only global warming but also, to take up but one example, the dissemination of microplastics, which may harm not only the corals and the fishes but also the ocean and those that live off it. The crisis is accelerating in the sense that its manifold and deleterious consequences are cascading and exacerbating. Far from being delimited and constant, the environmental crisis may elude not only our policy responses but also our comprehension, not unlike a rolling ball that begins its movement on a flat surface but slowly drifts downhill, all the while multiplying not only its speed but also itself, and suddenly drops over the edge. The metaphorical cliff may not exist, but its possibility cannot be wished away.

Why the systematic disjuncture between scientific figures and the actual numbers that supersede them, as in the successive IPCC reports? Or, more important, why the belated emergence of a conclusion? Only in 2007 did the IPCC manage to state that: "Warming of the climate system is unequivocal" (IPCC 2007). With the (brief) passage of time, what at first seemed to some a dangerously and irresponsibly alarmist screed turns out to be a modest and measured report (e.g., Broecker 1975). McKibben's *The End of Nature* reads like an arcadian tale by the time Wallace-Wells's book appeared. Why would the writers who sought to amplify the alarm understate, and not exaggerate, the peril?

No one wants to be the unhinged loudspeaker screaming doom. Most polemics are rhetorically unpersuasive and professionally undignified. The default tone of a responsible professional, whether a scientist or a journalist, is soft and steady, if not murky and muddled. Alarmism and doomism are for fanatics and vulgarians. Indeed, even McKibben and Wallace-Wells soften the blow of their jeremiads. The former reassured readers that "by the end of nature I do not mean the end of the world" (McKibben 1989, 7), whereas Wallace-Wells's book is strewn with random pronouncements of optimism, which he underscored in a later reconsideration (Wallace-Wells 2021). Would-be Cassandras don't want to be Cassandra, at least in Aeschylus's classic version, because they recall her descent into madness, not the irrefutable truth she tells.

Far from being an unalloyed good, scientific knowledge is a double-edged blade: at once shrouded and blunt. As the most prestigious and reliable mode of distributing information and insight about the world, best-selling books or scientific reports rely on authorized facts, figures, and forecasts, and pride themselves on their accuracy and even exactitude. Certainly, we cannot proceed without the scientists and their findings, but they are also slow and tentative, and least helpful precisely when the reality that the science seeks to analyze and explain is shifting rapidly. This is especially the case for big phenomena and large processes.

There are three ways in which modern science operates that perform a disservice to our understanding of the environmental crisis.

First, scientific prowess depends on the extensive division of intellectual labor, which retards the work of synthesis. In particular, the chasm between the study of nature and that of society prevents us from appreciating the hybrid character of natural degradation.

Second, scientific knowledge excels at falsification and analysis, but is less adept at macroscopic hypothesis generation and conceptual synthesis. Scientists mercilessly subject truth claims to the grindstone of falsification (and verification), but they are slow to assert and affirm the existence of a large or synthetic phenomenon. Put differently, scientists swiftly focus on the proverbial trees, but they are slower to recognize the forest. Their concreteness is usually not misplaced, but its weakness emerges in a moment of manifest emergency.

Finally, an epistemic virtue of modern science is its authority based on the scientific method. Mathematical modeling, methodological rigor, and other vaunted procedures legitimate knowledge claims, but the proximate arbiter is a community of scientists who rely on peer review and arrive at truths without coercion or distortion. The democratic process of knowledge construction makes modern science a triumph of the human spirit, but it takes time. It lacks the efficiency of diktat. Time is of the essence when reality is spinning out of control.

To overcome these shortcomings, we should not forget the inevitable entwinement of science and society and the elusive nature of science's self-proclaimed virtues. Most crucially, no one can avoid the essentially nonscientific character of our basic presuppositions. Science is an amazing achievement, but it offers neither omniscience nor omnipotence.

Division of Labor

Contemporary historians disagree on the nature of the Scientific Revolution—"There was no such thing as the Scientific Revolution," declares one authority (Shapin 1996, 1)—but the grand narrative of modernity posits a series of great transitions, such as from feudalism to capitalism, of which a critical one is the declension of religion and the ascension of science. As Pope rhapsodized, "Nature and Nature's Laws lay hid in Night. / God said, *Let Newton be!* And all was Light." The discoveries of Copernicus, Galileo, and others not only eclipsed Aristotelian and Christian natural philosophy but also "changed the character of men's habitual mental operations and the very texture of human life itself": a revolution that "outshines everything since the rise of Christianity," gushed an eminent historian (Butterfield 1950, viii). The promise of one true account of nature seized and served the scientific imagination.

The triumphalist view of the Scientific Revolution stresses the fusion of mathematical reasoning and systematic empiricism (and experimentation), and scientific knowledge thereby encompasses *inter alia* logical rigor (coherence and consistency), generalizability and universalizability, exactitude and certainty, and, for good measure, predictability (e.g., Galileo [1632] 1890, 129). As a contemporary authority declares, "laws of nature" are about "universality and necessity" (van Fraassen 1989, 1). Punctilious protocol and scrupulous operationalization enable replication and application: the stuff that makes a moonshot possible. In the early twenty-first century, most people associate science with objectivity, value neutrality (disinterestedness), skepticism, and the progressive and cumulative accumulation of knowledge: not so different from how it was characterized by the mid-twentieth-century sociology of science (Merton [1942] 1973).

I will have more to say about the popular understanding of science in the following chapter, but I focus here on the nature of scientific knowledge. Scholars disagree on exactly what makes science successful. Is it the scientific method or the logic of conjectures and refutations? They also question the unity or homogeneity of sciences. Is there a single principle or procedure, or a set thereof, that unites Newton's mechanics and Fabre's entomology or, for that matter, string theory and ornithology? The standard picture of science—the enshrinement of Newton and Einstein or the valorization of pure theory—is physics-centric and does a disservice to illuminating the workings of other scientific endeavors. The same statement can be made about the history of science and science studies. The single most influential book on the topic relies on evidence from the evolution of physics (Kuhn [1962] 2012). Whether we think of the prevalence of mathematical reasoning or systematic experimentation, the possibility of exactitude or predictability, it is descriptively the case that scientific fields and methods are diverse and disunified (Galison and Stump 1996). Indeed, there is no such thing as the scientific method, whether we comb through the writings from Plato to Mach or those from Darwin to Dewey (Laudan 1968, 38, Cowles 2020). From the messy reality of concrete practices, however, we abstract a cleansed set of procedures, which is tantamount to an act of violent reification.

Whatever the nature of science and the source of its success, no one would gainsay the salience of specialization. We don't need Smith or Durkheim to tell us about the importance of division of labor, which allows an individual or a laboratory to work collectively on a delimited sphere of inquiry. The enterprise of knowledge since the beginnings of the European university is coeval with disciplines, a systematic form of specialization (Kristeller 1974, 3). Scientific discipline avoids dilettantish speculations and embraces focused research: amateurism is out, professionalism is in. Rather than a few heroic geniuses, a community of scientists achieves small but tangible gains that add up to all that we know. Division of labor is critical not only for the modern factory, but also for the factory of modern knowledge.

The salience of division is not only a matter of social organization but also one of intellectual specialization. Not only is science emancipated from religion and other modes of prescientific knowledge, but it also declares independence at once from supernatural forces and humanity. In particular, the study of nature is severed from the study of culture or society, and thereby bifurcates reality into distinct systems (Latour 1991, cf. Whitehead 1920, 21). The separation may be as old as Socrates (Strauss 1970, 83), but the analytical divorce of natural science from human science is indisputable in the modern world. We are far away from seventeenth-century England where Boyle and Hobbes worked across the modern divide—Boyle's entanglements in politics or Hobbes's obsession with geometry are remembered only by erudites (Shapin and Schaffer 1985)—and we now remember Boyle as a natural scientist and Hobbes as a proto-social scientist, if they are recalled at all.

What is wrong with division of labor? Simply said, reality crosses the nature–culture boundary line, and the artificial separation obscures our grasp of it. A relevant variable or factor, if it should be on the other side of the divide, is disregarded. Thus we misrecognize what we think we know (Descola 2005). To take up a historiographical example, there was a long-standing debate on the "general crisis" of seventeenth-century Europe (Trevor-Roper 1959). What caused all the political upheavals and social discontent in that turbulent century? One hypothesis that was not mooted until recently was the Little Ice

Age, which impaired agriculture, placing inordinate stress on political and social life (Parker 2013, 684). Or consider the great question about the Rise of the West. Intellectual giants have slogged through archives and proffered one explanatory framework after another. Yet one neglected factor was the proximity and exploitation of coal that was crucial for the British and later the European industrial burst, in contradistinction to the situation in China and the rest of Asia (Pomeranz 2000, 68). Capitalism, in this outlook, is congenitally fossil capitalism (Malm 2016), which was an insight that was once popular before the efflorescence of the social sciences (Clark and Jacks 2007, 39). Parker and Pomeranz may be wrong, but few would deny after the Covid-19 pandemic that infectious diseases shape human history. It is bracing to realize, then, that the first English-language monograph on the 1918–20 Spanish flu that killed more than 50 million people around the world was published nearly sixty years after the scourge (Crosby 1976). Trauma alone cannot explain the long delay in making sense of the quintessential nature–culture fusion or hybrid: the natural *and* social history of infectious diseases.

The significance of nature to the study of the social world seems incontrovertible, but the relationship moves also in the other direction. The presumed autonomy of the natural realm and of natural laws justifies the exclusion of human beings in the study of nature. Yet it is not an accident that the idea of the laws of nature emerged from that of natural law in legal and social philosophy (Kelsen 1946, 263–64, Needham 1956, 518, d'Entrèves [1951] 1970, 22). To understand a pandemic, for instance, it is inadequate to investigate the genomic constitution of a pathogen or the physical vectors of dissemination. Human attitudes and behaviors are indispensable to their intensity or extension (Snowden 2019, de Waal 2021).

Nevertheless, social factors are belittled, if not ignored altogether. Biomedical research retains primacy, and the nature–culture hybrid called public health is marginalized. In concrete figures, for example, in 2020 about $12,530 per American were given to medicine but a mere $116 to public health (CMS.gov 2021, United Health Foundation 2022). Although these figures are inexact and are not strictly commensurable, the point about the vast disparity should be clear. Symptomatically, the Covid-19 pandemic led to an outsize hope for a cure-all vaccine, but even after its dissemination the salience of antiquarian practices, such as mask-wearing, handwashing, and social distancing were instrumental in minimizing the pandemic's devastation. To consider another example, Semmelweis's proposal that surgeons wash their hands before surgery may have eventually saved more lives than any biomedical innovation, but he was ignored at the time and remains buried in obscurity as we focus on pure science, as if biomedical research can expunge the social dimension of human beings (Nuland 2003, 183–84). To return to our main concern, the Anthropocene, a new epoch, is defined by anthropogenic destruction of nature (Crutzen 2002), though it is not accepted by the International Union of Geological Sciences as of 2024. However anthropocentric—it is surely part of the environmental problem that human beings are consistently, though understandably, obsessed with themselves and their place on Earth—the term at least points to an unprecedented crisis in the brief history of *Homo sapiens*, caused by human beings and their activities. Nature cannot escape humanity.

Division of labor runs deeper, however. The sovereign realms of nature and of culture are riven with fissions, which congeal into a series of autonomous and autarkic fiefdoms that we call disciplines. In the social sciences, major ones include those that study money and markets (economics), power and politics (political science), and what's left, such as family and crime (sociology), along with those that study other places (geography), other peoples (anthropology), and other times (history) (Wallerstein et al. 1996, 18–20). Anthropology, for example, is itself conventionally divided into four subfields: archaeology, biological anthropology, linguistic anthropology, and social anthropology. Social anthropology, according to a past authority, "examines societies 'without history', and cultures of an 'exotic' nature" (Nadel 1951, 6), but not nature or the environment until recently (cf. Kohn 2013). Although right-thinking contemporary social anthropologists would wail at the definition, the subfield has fissured into sub-domains and few can articulate what holds them together beyond their function as an institutional producer of doctorates that in turn serves as the labor market for disciplinary, and therefore disciplined and legitimate, professors (Abbott 2001, 140–41). The hypertrophy of higher education since World War II has increased the number of disciplines that answer to the demands of the state, society, and students (Jencks and Riesman 1968, Geiger 2019). Sociology, for instance, has undergone repeated fission to spawn independent disciplines such as criminology and gerontology, ethnic studies and gender studies. The more the merrier: almost all social scientists regard specialization as an unvarnished good.

Thus, one way to conceptualize the evolution of science is disciplinary differentiation or, in the language of Darwin's heirs, epistemic speciation. New disciplines emerge, not unlike new species (though the pace is faster). Just as older evolutionary biologists stressed biological reproduction as the arbiter of species survival and definition, scientific disciplines reproduce themselves by inward intercourse: publishing in disciplinary journals, attending disciplinary conferences, and working within discipline-defined departments. Scientific concourse across disciplines is often deemed pointless, so cross-disciplinary fertilization is often devalued.

Scientific disciplines also face organizational involution. In any large-scale enterprise, there is a powerful proclivity for a means to transmogrify into an end; a political party that espouses an ideal may subsume and even expunge it in favor of seizing and maintaining power (Ostrogorski 1902, 700). The search for truth in organized, professionalized science transmutes into a struggle for disciplinary efflorescence, boundary maintenance, and career advancement. The dynamic of organizational involution may temper, and possibly pervert, the quest for truth to an extrascientific desideratum.

The consequence of excess division of labor is hyper-specialization that fosters tunnel vision. As if burrowing a narrow, deep hole, mole-like investigators look forward—albeit at a very short distance—for the proverbial light at the end of the tunnel, but without much concern for matters in other directions: there can be no lights upwards or downwards or sideways. Myopia has its signal strengths, but striking shortcomings as well. In the biosphere, in contrast, specialization does not preclude inter-species communication and cooperation; symbiosis is common. This is rarely the case in the scientific sphere. "Specialists without spirit, sensualists without heart" is the implicit rallying cry of scientists, without the gloss: "this nullity imagines that it has attained a level of civilization never before achieved" (Weber [1904–5] 1930, 182). The danger of unbalanced specialization—the proliferation of professional fools—is passed over in silence (Ortega y Gasset [1930] 1932, 112–14). The iron diktat of scientific progress is "more research is needed" in one's chosen specialization, but everything else is "not my field" and therefore beyond one's purview: my way to truth or your highway of irrelevance.

Excessive division of scientific labor obfuscates a world that is *not* divided into disciplines. The phenomenon of power permeates every sphere of social life, for instance. If we study the market, then we may see phenomena that will be classified under political science (state policy and intervention), criminology (white-collar crime), or anthropology (trade with or migration from the Global South). Abstract division maps poorly onto concrete reality. The trend has been to cover the entirety of the social world in one discipline. Hence, psychology exists within economics (economic psychology), political science (political psychology), and sociology (social psychology), none of which communicates much with each other. Similarly, in addition to the discipline called history, there are economic history, political history, and historical sociology, but these subfields rarely interact. Disciplines maintain their tenuous hold on reality not by engaging in interdisciplinary communication and collaboration, but rather by expanding their own sphere beyond the original remit. Indeed, we are awash with attempts at intellectual imperialism—as with human history, the histories of scientific fields are littered with *disjecta membra*—and little to show for all the expansionary zeal. There is no reason to abjure specialization and division of labor, but the work of integration and synthesis remains inchoate and incomplete.

One remedy would be to designate a group to engage in the work of synthesis, as in Comte's ([1830] 1975, 88) vision for sociology, which is far from the existing reality of that discipline. A more viable response to disciplinary division and its discontents is *interdisciplinarity*. The call to bring together what has been torn asunder is coeval with the history of human knowledge. In matters of the nature–culture divide, some Renaissance scholars had already sought to reintegrate what we would call science and religion as *prisca theologia* (Yates 1964, 17–18, 393), and others crossed the rift between the study of nature and that of humanity (Yates 1972, 292). Transdisciplinarity is as old as any inter-species of knowledge (Hugh of St. Victor [1176–77] 1961, 890; cf. Collingwood 1924). But the call for interdisciplinarity and transdisciplinarity remains stillborn in spite of its rhetorical fanfare.

Division of labor promotes scientific advances, but it may also retard and occasionally stifle them. The problem becomes acute for a finding that does not fit the conventional grid of knowledge, in a newfangled interdisciplinary endeavor. Hybrid fields of inquiry, if only because of their newness, receive less attention and support. Herein lies the institutional and intellectual cause of the slow development of environmental science. The global environmental crisis, which entails synthesizing complex, heterogeneous phenomena that span the spectrum of human knowledge, is elusive in and resistant to the conventional division of scientific work. Recall the crux of global warming: carbon dioxide emission and the greenhouse effect. Although the lineaments of its workings were identified as early as 1824 and replicated in a laboratory experiment by 1859, it was only with Revelle and Suess's 1957 paper that the problem received attention in the post–World War II period. Yet their work did not fit within the extant disciplines and fell between the disciplinary silos of knowledge. What could have been central to the newfangled field of environmental science was rendered moot by its absence until the 1970s and 1980s (Oreskes 2021, 491–92). To be sure, there were other factors at work, such as funding. For example, it was precisely when US Navy funding declined in the 1980s that other research areas, including carbon dioxide emission, came to the fore instead of those that the navy had deemed significant in the Cold War decades (Oreskes 2021, 491–92).

In fields that hanker for scientific legitimacy, any significant departure from convention, regardless of whether it strikes the investigator as an unvarnished truth, may be moderated by a desire for respectability. At a minimum, additional evidence must be compiled and the rhetoric of presentation hedged to cocoon a deviant idea in the cloth of conventional wisdom. This is one cause of the belated character of IPCC warnings. Rather than dwelling in the interstitial space of interdisciplinarity, then, the natural temptation is to erect a new discipline. The Association for Environmental Studies and Sciences was founded only in 1983 (most social science disciplines, in contrast, were established in the late nineteenth century—1885, in the case of the American Economic Association). From the obscurity of liminality to the spotlight of centrality, disciplinarity reigns.

Finally, there is another significant divide. The valorization of pure or basic research—made possible in part by the rapid expansion of government funding and higher education in the post–World War II decades—obviated the desire to reach a wider audience. Earth scientists, for instance, routinely wrote popular works before World War II, but they focused on professional, intradisciplinary communication thereafter (Oreskes 2021, 482–83). In digging deeper into each hole, the promised work of synthesizing disparate discoveries is not built into each discipline or across disciplines. Strictly speaking, work of synthesis is stillborn and outsourced to outsiders, usually journalists, as we have forgotten the wisdom of past thinkers: "For even as it is better to enlighten than merely to shine, so is it better to give to others the fruits of one's contemplation than merely to contemplate" (Thomas Aquinas [1485] 2012, 817). We are, then, left without a vade mecum—an intellectual guide—precisely when we are confronted by complex, confusing, and changing reality.

Scientific knowledge—including social-scientific knowledge—developed at cross purposes with the complex and heterogeneous reality of the environmental crisis. Specialization accelerated at the cost of synthesis, and we were therefore slow to realize the totality of what scientists had been chipping away at in their isolated frontiers. In most matters, this would not be a serious issue. In the case of the global environmental crisis, however, its pace renders the delay devastating.

Differentiation and Integration

An indisputable strength of scientific knowledge is its empirical orientation. Facilitated by division of labor, scientists work on defined, delimited areas of inquiry, which in turn eases the task of verification or falsification. The obverse of scientific rigor—and related to specialization that may occlude a picture of the whole—is reticence toward a proposition that is difficult to verify or falsify. Thus, a crucial concept, such as energy, can remain elusive and resist rigorous definition (Smil 2022, 23). A novel phenomenon, especially something large but shifting in character, would be another case in point. It is easier for scientists to ascertain temperature increase in a particular area and how that affects corn growth and fecundity, for instance, than to assess the seemingly nebulous idea of global climate change. Analysis is constitutive of scientific thinking, but synthesis is a struggle. Science excels under the condition of *differentiation*, but falters at times in situations where *integration* is crucial. The will to ignorance of the whole is an integral feature of the scientific-knowledge enterprise and exacerbates the dysfunctions of division of labor.

It is a truism to say that modern science's beauty and majesty stem from its critical power, most notably its corrigibility. Scientists seek to verify their hypotheses rather than merely dictating them as fiats or, worse, confounding them. As a noted physicist remarked, he "would like to be understood in an honest way rather than in a vague way" (Feynman 1965, 13). Scientists also learn from mistakes as they attempt to refute received claims. Rather than relying on dogma or authority, then, scientific knowledge is buttressed by the twin processes of verification and falsification. Never mind that we have not verified the verification principle or that falsificationism is not falsifiable. Scientists routinely come to agreements that are supported by logic and evidence, and whoever fails to squeeze through the strait gate cannot legitimately assert the status of scientific knowledge.

"Prove it" is the mantra of scientific authority, then, however philosophically dubious its procedure may be. Proof is also laborious. Whitehead and Russell's (1912, 86) painstaking effort to erect a foundation of mathematical knowledge required over a thousand pages to show that 1+1 = 2. Yet the tragic reality is that their mammoth endeavor rested on an untenable philosophical premise. Gödel's (1931) second incompleteness theorem—that a formal system cannot prove its own consistency—swept away the epistemological rug upon which the two philosophical giants stood. The solid ground of certainty proved illusory, shorn of the foundation of its truth claims (Kline

1980, 352). Wittgenstein ([1956] 1978, 14–54), who criticized his mentor Russell's edifice, famously did not believe in the proposition that "There is no hippopotamus in this room at present" (Russell 1951, 297). It is notoriously challenging to prove absence. In the dialectic of proofs and refutations, however, we discover the virtues of science— if mathematics can be considered a science for the sake of my argument—as nondogmatic and democratic in the sense of being open to challenges (Lakatos 1976, 4–5). What interests me here is not metamathematics or scientific methodology but rather the sort of propositions that face the mill of proofs and refutations. The vaunted virtues of scientific knowledge—specialization, reductionism, exactitude, and so on—tend to look down, so to speak, to the concrete ground, not to gaze abstractly upwards. To alter Whitehead's fallacy of misplaced concreteness, modern science is inept at noticing a hippopotamus in the room when it comes in the form of big structures and large processes, such as the global environmental crisis.

Consider something as basic as the foundational ontological structure of biology. After Stanier and van Niel (1962, 18) sought to rectify "the most remarkable feature of this extraordinary situation [which] is that there has been so little argument" about the discovery of the proliferation of bacteria, the tree of life, according to modern biology, was said to consist of two kingdoms: prokaryotes and eukaryotes (single-celled and multi-cellular organisms, respectively). Woese and Fox's (1977, 5089) proposal for a third "urkingdom" of "archaebacteria" upended the orthodoxy (later called "domains" and "Archaea," respectively [Woese, Kandler, and Wheelis 1990, 4578]). Although Archaea were the sole extant organisms for billions of years and still dominate life forms on Earth, recognition occurred only in the past half-century, and scientific agreement took a few more decades. Indeed, intellectual resistance was fierce and ferocious (Mayr 1998, 9723, cf. Sapp 2009, 269–81). If nothing else, the centrality of Archaea—though not surprising from the decentered standpoint of evolutionary theory—goes against the grain of anthropocentrism. The preponderant presence of Archaea and Bacteria in the biosphere sits uneasily with the notion, popular at least since the Italian Renaissance, of "man the measure" of all things. The central figure, Woese, became "bitter" because of the belated recognition of his (and his colleagues') work (Quammen 2018, 9). Yet the three-domain model stands now as the orthodox tree of life. It is possible to recast this story as yet another triumph of scientific progress, but it can also be seen as a case study in the difficulty of *integration* or synthetic knowledge.

Modern science has a problem with synthetic, integrative knowledge. An example from the social sciences may elucidate my point. It would be easy to lambaste the notion of capitalism because of its abstract intangibility, but few social scientists today would hesitate to use the term as it captures, *faute de mieux*, the social reality of our political–economic system and culture. Since it was popularized, though not coined, by Sombart (1902), the term has steadily gained acceptance, though rejected by some for its Marxist overtones (Sonenscher 2022, 24–25). To be sure, most people who use the term capitalism do not regard it as a matter of logical and necessary truth, but employ it for its convenience and utility. Hence, it is much closer to being a model or an ideal type; we do not seek apodictic certainty or logical coherence but rather depend on it as a shorthand to visualize phenomena that are intangible and imprecise. Furthermore, like all important social-scientific terms, such as democracy and equality, it is essentially contested (Gallie 1964, 174–75); people hold conflicting, and at times contradictory, definitions and understandings of the term. In brief, it is a mess. Yet we would have trouble doing social science—and indeed even talking about political economy or sociology—without abstract concepts of the state, capitalism, society, and so on that perforce lack conceptual consensus and precision. Some scientists may regard a social-scientific concept of this sort as something akin to ether: an unnecessary idea and wrong to boot. Modern science, after all, prospered precisely by eschewing difficult problems, such as causality, and consigning them to the realm of the metaphysical (Crombie [1952–59] 1961, 319–20). Yet what term would they offer to make sense of our political–economic system? Even at its lower reaches, our world is replete with bewildering and perplexing, enigmatic and unfathomable objects and hyperobjects.

The culture of modern science revels in the task of *differentiation*. All its signal virtues, such as precision and rigor, can flourish in subjecting hypotheses to close examination by dint of specialization. The tendency is to inspect microscopic, grounded, and delimited areas of inquiry. In contrast, the work of *integration* eschews exactitude and confronts elasticity. There is a conceptual leap based on faith or imagination, whether to demarcate organismic domains or capture, in a simple but inexact concept, something about our political economy. The challenge is multiplied when a phenomenon may be unstable and incoherent, large and intangible (cf. Morton 2010, 130–35). Needless to say, scientists embrace new objects and hyperobjects, classification schemes and conceptual renovations, over time, but they remain by training and temperament reluctant to do so. As we saw in the case of Archaea, preeminent scientists resisted recognizing this third group of organisms. A new generation of scientists, who learned of Archaea in textbooks, merely accept them as part of science, part of nature. One may celebrate this as the structure of Scientific Revolution, but the banal reality is that, as Planck (1948, 22) quipped: "The truth never triumphs, its opponents die."

What self-respecting scientist would propose what is tantamount to a metaphysical hypothesis, such as the global environmental crisis? It does not take an analytical philosopher to slay all three of these ill-defined terms—"global," "environmental," and "crisis"—and the idea seems safely distant from the realm of rigorous empirical inquiry. The motivating idea of ecology or environmental studies is that "everything is connected"—or, as Carson ([1962] 2018, 51) declared: "in nature nothing exists alone"—but this challenges the favored scientific modus operandi of differentiation and analysis. A generation of scientists for whom the global environmental crisis is as obvious as the atmosphere they breathe had to arise before the crisis became scientific common sense and a field worthy of scrutiny. Even in a synthetic discipline par excellence—philosophy—neither ecology nor the environment merited an entry in the 2006 ten-volume *Encyclopedia of Philosophy*, which provided only two short entries for "environmental aesthetics" and "environmental ethics" (Borchert [1967] 2006).

In short, there is an epistemological obstacle to scientific knowledge. Its analytical prowess supersedes—and in many ways is dependent upon—its synthetic poverty. The will to ignorance of the whole retards the emergence and establishment of a useful, synthetic idea, such as ecology or the global environmental crisis. This is a problem, again, primarily because of the accelerating, escalating threat.

The Legitimation of Knowledge Production

The epistemic superiority of scientific knowledge lies in its democratic constitution and corrigibility. Rather than rely on external powers, whether divine or human, that dictate truths, scientific endeavor seeks to legitimate its knowledge claims. Logical ratiocination and mathematical reasoning, or systematic observation and experimentation, play their part, but hypotheses and hunches become scientific knowledge because scientific peers become convinced of their plausibility. Although external validations—a perch at a powerful university or prestigious prizes—play their part, scientific knowledge is less about the scientist and more about the knowledge itself. What they know, not who they are or whom they know, is, or should be, the key. Founded on neither sacred nor secular authority, nor even charismatic genius, scientific knowledge must survive tests and trials. Systematic critical scrutiny validates and valorizes scientific knowledge. It is only after lengthy tribulation that consensus crystallizes into textbook verities. Devoid of distortion or deception, we trust in science because of its procedural superiority (Kuhn [1962] 2012, Oreskes 2019, 19). Organized skepticism legitimates scientific knowledge.

Peer review plays a crucial role in certifying scientific knowledge. Credentialed and proven investigators adjudicate the propriety of logic, method, and substance as supplicants claim new knowledge or falsification of the old prior to publication. Peer review is the method of quality control that separates the wheat from the chaff.

In spite of the vaunted place of peer review in scientific-knowledge generation and legitimation, it remains a curiously underexamined practice (Shatz 2004, 4). As much as we celebrate its centrality in the workings of contemporary science—and scientists falsely give it a long lineage, usually to the seventeenth century or the time of the Scientific Revolution—we blush to learn that *Science* and the *Journal of the American Medical Association* did not adopt the practice until World War II (Spier 2002, 358). Predecessors to modern peer review abound, but its institutionalization occurred only in the mid-twentieth century. Peer review arrived quietly but swiftly, becoming synonymous with science itself, but its seeming permanence and naturalness strike a discordant note. None of Einstein's great articles, for example, were peer-reviewed. As the redoubtable Planck—who edited the most prestigious physics journal at the time, *Annalen der Physik*—proclaimed, the editor must have a "completely 'free hand' to make decisions as they went along" to achieve "the highest editorial principle" (Jungnickel and McCormmach 1986, 310). Indeed, all five pathbreaking papers that Einstein published in 1905, including the iconic one that introduced the most famous equation in the world, appeared in *Annalen der Physik* (Rigden 2005). Given that early twentieth-century physics—surely a time of power and glory—fared well without peer review, we may well wonder why so many scientists believe in its centrality and necessity.

To the extent that researchers have critically examined the practice of peer review, the collected litany of complaints is long and sobering. The process is labor intensive and requires an enormous expenditure of intellectual energy, but there are well-documented and widely acknowledged abuses, biases, and corruption as scientists game the system in their quest for fame and funding (Ferguson, Marcus, and Oransky 2014, 481). We know that the process is not effective in detecting errors or frauds (Stroebe, Postmes, and Spears 2012, 683). Professional reputation and prominence sway the process, such that the author(s), rather than the paper(s), determine publication and influence (Huber et al. 2022). The limited number of acceptances for publication in established journals means that the process is tantamount to a lottery, for many worthy articles are rejected. There is also the fundamental paradox of instituting a practice that restricts the dissemination of ideas in the proverbial marketplace, while in the age of digital publications there are fewer compelling reasons, such as space constraints or publishing costs (to be sure, neither is space infinite nor are costs zero), to restrict the number of acceptances. More troubling, agreements among referees rarely exceed one-third (Cicchetti 1991, 126–27). Breakthrough ideas become enmeshed in the seemingly holeless net of peer review that appears eager to eliminate deviant knowledge (Smith 2006, 178). Indeed, a perverse incentive for publish-or-perish scientists may be to thwart innovation (cf. Horrobin 1990). Hence, scientists who examine peer review often end with pejorative conclusions, such as that it is "an untested process with uncertain outcomes" (Jefferson, Anderson, Wagner, and Davidoff 2002, 2785). The search for truth and the process of peer review seem to diverge (Ritchie 2020). Yet few contemporary scientists would question its primacy and legitimacy.

Peer review also takes time: scientists referee papers and editors make judgments based on them. Anticipating potential criticisms, scientists temper controversial claims and expend efforts to rebut novel ideas. The quest for respectability—a crucial currency in any community—dampens new and potentially controversial claims. Science moves slowly and carefully, and remains antithetical to tabloid journalistic virtues of sensationalism and scandalmongering.

Second, and more important, even if peer review operates well in an established and well-populated field, the same cannot be assumed for a new area with few active scientists. Dubbed "normal science," routinized problems—making an incremental change in an established subfield—are readily recognized and rendered capable of competent judgment. However, a new idea that upsets the conventional wisdom or contributes to a new field does not have the preexisting audience and ready reserve of reviewers. Herein lies another challenge for an inchoate field of inquiry, such as environmental studies.

The challenge is in fact overdetermined. Not only is environmental science new, but it is also a poor fit in the regnant division of intellectual labor. As a transdisciplinary endeavor, it fails to conform to received ideals of specialization. Indeed, the entire enterprise carries a whiff of illegitimacy. In brief, the slowness of the peer review process redounds with additional obstacles for validating the very idea of the environmental crisis.

The Entwinement of Science and Society

Division of labor and specialization, resistance to big structures and large processes, and the slow process of knowledge legitimation contribute to the belated character of scientific knowledge. As part of a new, controversial, and interdisciplinary field, studies on the global environmental crisis face additional hurdles that retard their acceptance as consensual truth and dissemination to the wider world. Moreover, claims about the global environmental crisis confront obstacles intimately intertwined with the presuppositions of modern science, namely, beliefs in its autonomy, objectivity, and disinterestedness. The captains of erudition proclaim that "the scientific spirit" is "the culture of the idle curiosity" (Veblen 1918, 176).

Modern science champions the values of neutrality, purporting that scientific knowledge owes nothing to money, power, fame, or other extrinsic rewards. A interdisciplinary presupposition is that scientific research is driven by curiosity and is therefore pure, knowledge for its own sake, devoid of interest and passion. Partisanship is a poor rhetorical strategy to convince others, and knowledge claims would be rendered more cogent by masking a prior value commitment or vested interest. The ploy of being objective, neutral, pure, and value-free makes a claim all but irresistible.

What gives science epistemic authority? It is common enough across culture and history that the monopoly of truth coincides with the monopoly of power: "All aristocrats hold the fundamental conviction that the common people are liars. 'We truthful ones'—that is what the ancient Greek nobility called themselves" (Nietzsche [1886] 1998, 154). In early modern England, the question was not whether knowledge is pure, but rather *who* could be trusted to carry out responsible and credible investigation. The answer, not surprisingly, was the figure of the gentleman, who is endowed with the virtues of civility and gentility, honor and trust, and therefore possesses credibility and reliability (Shapin 1993, 66–68). Our modern, democratic temperament recoils at past forms of epistemic authority. Truth should be true no matter who utters it. Instead of status, modern scientists are said to be disinterested, save for their commitments to rules and procedures that generate truth. They labor to uncover the workings of nature and are motivated not so much by fame or fortune but by idle curiosity, the will to knowledge.

Not only is idle curiosity a garden-variety human impulse, but it also often runs against the grain of modern science. Ancient Greeks, for instance, cultivated idle curiosity but this led them to speculations and aversion to experimentation (Sambursky [1956] 1987, xiii, 228–31). As Hume ([1739–40] 2007, 286) said of curiosity, "love of truth … was the first source of all our enquiries." Yet Housman is not altogether wrong to suggest that "the faintest of all human passions is the love of truth" (Manilii [30–40 CE] 1903, xliii). There is no point in denying that idle curiosity has been the springboard of scientific discoveries, and much else besides in human endeavor, but perhaps the most important element of idle curiosity is its antiauthoritarian potential. We shouldn't be surprised, then, that curiosity raised suspicions in early modern England, associated as it was with insubordination, insolence, and impiety (Benedict 2001, 25). Its effectiveness depends on its institutionalization: it should not be just an individual virtue, but also a systemic one.

There is something sinister about all the talk of idle curiosity that fuels the ideology of pure science. The vision of science free of fear or favor becomes crucial in the post–World War II period precisely when scientists in free society, especially in the United States, sought to combat totalitarian enemies of the right and the left. Succumbing to neither racism nor Lysenkoism, scientists in democracies hoped to pursue *pure* science in an endeavor that replicates the values of capitalist democracy: science functions as a free marketplace of ideas, devoid of coercion or deception (cf. Merton [1942] 1973, 276–78).

Just as a truly free market is a limiting condition—what does it mean to talk about the free market when state intervention is pervasive, and oligopolies are so common?—the idea of pure inquiry rarely manifests itself. Pace Hardy's apologia for pure mathematics, it is not an accident that the development of pure mathematics has been intimately intertwined with political–economic interests and technological–industrial concerns (Kline 1953, 7, 11–12). Is it possible to disassociate the study of trajectories from the military interest in projectiles? Surely Needham (1956, 542–43) was not wrong to highlight the roles of commercial and capitalist interests in the rise of modern science in the West.

Disinterestedness is a ruse because even idle curiosity presupposes a wide array of social and personal factors that amount to a vested interest, however devoid of desire for fame or fortune, honor or influence, it may be. And questions that dominate a particular place and time rarely emerge without brushes with real-world concerns. Given the infinite number of problems one could work on, there is an implicit value choice in any scientific investigation. It may not profit a woman to trace the natural history of a pulsar, but her devotion entails a human decision made in a particular milieu at a particular time, and therefore discloses an implicit—at times explicit—value commitment.

In the early 2020s, few would question the carcinogenic character of cigarette smoking, and we also know about the half-century of intense lobbying and funding efforts to refute oncology research that revealed the indubitable connection between cigarette smoking and cancer and sow seeds of confusion about scientific research (Proctor 2012). Although the evidence that links cigarette smoking to cancer is about as conclusive as any scientific finding can be, there were scientists who, either out of conviction or desire for funding, kept the controversy alive and thereby provided ammunition, studded with scientific credibility, for the tobacco industry and its products. Individual variations (e.g., a chain smoker who dies in old age, blessedly free from lung cancer or other diseases), the long gap between smoking and the onset of illness, and other biosocial conditions provided plenty of space to poke holes in the claim of causality that was far short of one-to-one determinism. As I elaborate in the following chapter, the public has embraced a scientific worldview that depends on determinism when the world works probabilistically. Full determinism—the elusive goal of certainty—is unrealizable. From fighting factual evidence to promoting alternative viewpoints, lobbyists and scientists served as vassals to the knights of the powerful and profitable tobacco industry (Michaels 2020).

In a more abstract language, the democratic character of scientific knowledge offers room and cushion for free inquiry, however perverse and mercenary it may be. An inevitable consequence is that one or another credentialed figure can be mobilized to generate doubts on a seeming certainty, and that makes the addiction just that much more palatable for consumers.

What the tobacco lobby has done to smoke the link between cigarettes and cancer is replicated by the coal industry and its allies to gaslight the causal path from fossil-fuel burning and other human-generated greenhouse gas effects to climate change and global warming (Oreskes and Conway 2010, 7–9). Earnest, ideological, or merely mercenary, some eminent scientists sought to upset the foundations of environmental science and cast doubt among politicians, policymakers, and the public (Mann and Toles 2016).

Science does not operate in a vacuum. From funding to publishing, political and ideological struggles are constant. Some climate denialists engage in outright slander and misrepresentation, while others merely exercise the personal and organized skepticism that is part and parcel of scientific research. When social scientists, such as economists, become involved, they bring their hobbyhorses—discounting and externality—to stress the enormous cost of preventing global warming (Oreskes and Conway 2010, 179–80). Even if most scientists desist from mercenary mudslinging, the pervasive impact of paymasters is difficult to deny. In an exhaustive study of post–World War II oceanography, Oreskes (2021, 501) concludes: "Oceanographers shared or internalized the values of their Navy patrons." Sporting a mask of neutrality, scientists forget that their mask was shaped by the paymaster, and soon enough the mask becomes the man.

There is a deeper irony in that scientists became dependent on government funding precisely when the call for pure research became dominant. The ideal of pure science was hegemonic when the Cold War appropriation of defense spending fueled so much natural science research in the post–World War II era (Oreskes 2021, 470). Whether regarding the development of nuclear weapons or innovation in military technology, the front of disinterestedness crumbled against the threat of nuclear annihilation or protests against the Vietnam War (Jacobsen 2015, 215–18). Amid accusations of warmongering and other challenges to the self-subscribed image of disinterestedness, it should not be surprising that discombobulated claims were clutched tighter and transmogrified into visions of purity and neutrality.

Value neutrality is an impossible ideal in scientific investigations and debates about the global environmental crisis. Even if the majority of scientists subscribe to the idea of pure science—neutrality and disinterestedness—the existence of an interested party undermines the state of play. Repetitive and reflexive calls for objectivity and neutrality do little to block blatantly ideological pressures and play into the hands of the forces that would deny patiently accumulated scientific evidence.

What mires environmental studies in sticky questions of objectivity and neutrality is precisely its hybridity that impinges on both the natural and the social. Put simply, environmental scientists may not care about the spreadsheet of a coal company, but the latter cares deeply about the former and their findings. The pursuit of profit renders scientific knowledge a matter of contention that brings into the presumptive pure world an array of impurities, be it profit or political ideology. The will to truth clashes with the will to gain, the will to power.

In other words, scientists may be motivated by curiosity, but they are also entwined in the inescapable contexts of money and power, fame and honor, and credibility and respectability. There is a profound ideological presupposition among modern scientists that they are motivated above all by curiosity. As much as they may be tempted by money, power, and fame, almost all would abjure these worldly motivations. The widespread embrace of the ideal of pure science— the quest for truth or the valorization of basic research—appears as a benign, universal ideal, but there is no denying that historical and social circumstances have attracted many of the finest minds to one area of inquiry over another, whether theoretical physics in the early twentieth century or genomics in the early twenty-first century. In the process, new interdisciplinary inquiries are marginalized. The struggle for respectability works to temper controversial claims. Incendiary findings are doused with a shower of modesty. All these forces vitiated research on the global environmental catastrophe and its dissemination to the larger scientific community and, over time, to the larger public.

Nonscientific Foundations of Science

A deeper strand in the interlacing of science and society is that nonscientific presuppositions are inescapable.

What makes scientific knowledge attractive and robust is its universality. Scientific truth appears true across time and culture; the law of gravity operated billions of years ago as it does today, and in the United States and China or, for that matter, on Mars. However, the law of gravity, in its classical articulation, is *not* true for all times and places (Carroll 2022, 97). There is nothing like the twin developments of modern physics—theories of relativity and quantum mechanics—to destroy comforting illusions of a certain, stable, and knowable universe (Maudlin 2011). Without denying Newton's intellectual achievement, we should also not forget that there are limits to generalization of even the most powerful scientific laws and theories. No knowledge is possible without presuppositions, but presuppositions are ultimately beyond conclusive, endogenous proof.

Consider a commonsense example, that something—a person—cannot be in two places at the same time. If a suspect can prove that she was far away at the time of an alleged murder, then contemporary jurors would exonerate her of the crime. Yet during the Salem witch trials, a vision of supernatural visitations imperiled the accused. Spectral evidence—though we moderns would reject it—was a powerful, albeit contested and ultimately rejected, proof in court in seventeenth-century New England (Boyer and Nissenbaum 1974, 16–18). It would be simple to dismiss the mindset of the benighted Puritans, but contemporary physicists assert the possibility of a particle being in two places at the same time. Most self-regarding enlightened readers would be loath to dismiss scientific knowledge, however outlandish, presented as it is by credentialed intellects with mathematical equations and mysterious evidence. My intention is not to adjudicate the legal reasoning of early modern Puritans or to embrace the fantastic notion of quantum superposition. Rather, it is to assert not only the centrality of presuppositions but also their mutability.

A powerful presupposition of universality is the essential stability or order of the universe. As Whitehead (1929, 6) declared, "There can be no living science unless there is a widespread instinctive conviction in the existence of an *Order of Things*, and in particular of an *Order of Nature*." Along with the belief in an omnipotent maker, the presumption of order may be useful in devising abstract mathematical models that would be true across space and time. Still, the universe—and its constituent parts—may *not* be stable. The persistence of a phenomenon renders culture as nature, defining the very horizon of thought. To take one example, Kremlinologists assumed that the Soviet Union would last, perhaps not forever, but surely for a long time, such that few, if any, prognosticated its demise. Everything was forever until it was no longer (Yurchak 2005). It would not have occurred to the seventeenth-century scientific revolutionaries that planet Earth would be subject to anthropogenic devastation.

A similar presupposition is nature's simplicity. As Newton observed, "It is the perfection of God's works that they are all done with the greatest simplicity. He is the God of order and not of consusion [*sic*]. And therefore as they that would understand the frame of the world must indeavour to reduce their knowledge to all possible simplicity" (Manuel 1974, 48–49). Simplicity and order—and the existence of God the creator—are, of course, assumptions.

What is true for the universe is also true for political discourse or scientific knowledge. However expedient and reasonable, a moderate position between two extremes is not necessarily correct. At the height of the abolitionist debate in the antebellum United States, it may have coddled the *amour propre* of moderates to seek a balance between explicit rationalization of slavery and radical abolitionism; few today would resist the rightness of one extreme position. To accept Solomon's judgment—seemingly a moderate, balanced solution—at face value is precisely the proof of falsity and barbarity. That is, value commitment, not objectivity and neutrality, opens the door to wisdom. Ruptures in scientific convention—also known as scientific revolutions—generate passionate disagreements. To advocate a new, and seemingly extreme, position cannot be the product of extensive experimentation and observation or of objective and neutral deliberation. A new perspective needs to be justified and legitimated over time, but its initial articulation does not stem from the grinding of the scientific method.

In spite of considerable achievements, then, scientific knowledge is essentially limited. In denying its imperfection, some become arrogant and extol science at the expense of other nonscientific forms of knowledge. Yet science is not the only valid enterprise of knowledge, and it is neither omniscient nor omnipotent. Our knowledge of the global environmental catastrophe, like our knowledge of good and evil, does not stem from science.

The idea of environmental catastrophe rests in the realm of extra-scientific knowledge, proffered by scientists and nonscientists. It has generated a great deal of scientific knowledge, but we should not engage in an anachronistic act of consigning it to the routine exercise of conventional science.

Conclusion

Scientific knowledge is the most rigorous and reliable form of knowledge we have. Yet it does not allow us to achieve omniscience or omnipotence. It has contributed to the destruction of nature, but it has only belatedly enabled us to comprehend the environmental crisis. We should not and cannot reject scientific knowledge, but it is insufficient in and of itself to deal with the disaster that it has helped to precipitate.

2
FATE AND HUBRIS

In October 2019 *Global Health Security Index* assessed "the state of international capability for preventing, detecting, and rapidly responding to epidemic and pandemic threats" (Nuclear Threat Initiative 2019, 7). It ranked nearly two hundred countries on their state of preparedness, and the United States came in first, followed by the United Kingdom (Nuclear Threat Initiative 2019, 20).

Two months later, the first cases of Covid-19 were reported in Wuhan, and by February 2020 the disease had circled the world. Some speculated that Chinese scientists engineered SARS-CoV-2 and its leakage might have unleashed Covid-19 (Ridley and Chan 2021); very few noted that an artificial concoction would not gainsay its root as a zoonotic disease, which is engendered by the loss of wild habitats and the consequent spillover of a pathogen from an animal species to *Homo sapiens* (Garrett 1994). In other words, Covid-19 was part and parcel of the ongoing global environmental catastrophe.

Be that as it may, by March 2020, when the World Health Organization declared a state of pandemic, the world slowed down and nearly came to a halt from travel bans and lockdowns. By the end of the year, it appeared that far from being the best-prepared countries, the United States and the United Kingdom were among the least ready, if we measure readiness by mortality figures.

Here I am less interested in the lack of attention paid to the pandemic's linkage to natural destruction, and more intrigued by the ranking. How could a panel of distinguished scientists and practitioners get it so wrong? One source is the neglect of the all-important political factor, which is in line with the chasm between the study of nature and that of society discussed in the previous chapter. But there is more to the situation than outright political interference or institutional flaws and failures.

DOI: 10.4324/9781003539407-2

Power of the entrenched mindset—call it worldview—can be seen in the widespread resistance to vaccination, which became available by 2021. If Operation Warp Speed reflected the positive face of the power of science, then antivaxxers were heretics who embraced irrationality and unreason against the religion of science and technology. Most analysts are wont to ascribe an antirational and antiscientific mindset to antivaxxers or climate change denialists and blame one or another ideology as the source, be it right-wing populism or Christian fundamentalism (e.g., Carpiano et al. 2023). Despite the impact of these ideologies, we are not passive recipients of ideas from above, whether from charismatic media personalities or theatrical political leaders. Far from dwelling in the realm that castigates science and technology, attention-starved politicians and antivaxxers share the regnant worldview of science and technology. It is precisely because they believe in the promise of techno-scientific knowledge that Operation Warp Speed was conceived and funded. The scientific worldview presents scientific knowledge and technological application as something close to religion and magic. The vaccine, in this instance, was far from definitive and perfect, which provided fuel to antivaccination sentiments and movements. The scientific worldview deviates from the contemporary practice of science and technology, which operates in the realm of provisional truths and probabilistic results, rather than definitive truths and deterministic outcomes.

The Scientific Revolution promised certain knowledge. Being exact and predictable—armed with the assurance of mathematical equations—it seemed to offer one and only one true answer to a series of questions about nature. Rather than by deciphering a sacred text, mathematical formulae, systematic observation, and rigorous experimentation would yield the right answer. The monotheistic mindset, though changing the book it consulted, remained robust. What interests me here is less what seventeenth-century natural philosophers believed in, and more what their boosters and successors today profess: how the philosophical presuppositions of the Scientific Revolution came to be routinized and reified as a belief system. Disseminated in science textbooks and popular accounts, *scientism* is the regnant cosmology and worldview of the contemporary mindset in the Global North.

The scientific worldview, or scientism, is a simplified articulation of the Scientific Revolution and its legacy. In this line of thinking, science provides exact, deterministic knowledge that allows intellectual and material mastery over nature. Contemporary scientific knowledge proceeds, in contrast, with the presumption that the universe is probabilistic or stochastic. The chaos, contingency, and complexity of nature are elided in favor of the fairy-tale world of order, necessity, and simplicity in scientism. Exemplified by Newtonian mechanics, scientism stresses scientific knowledge as exact and certain, precise and predictable. In turn it fuels applications, or technology, that make the ancient seven wonders of the world seem child's play in comparison. The scientific worldview, moreover, perpetuates the rift between nature and culture, and sustains the belief that the workings of nature (or divine or supernatural forces) are matters of *fortune*, rather than matters for human intervention and therefore questions of social *justice* (Shklar 1990). Finally, scientism overshoots its purview and serves as a hubristic ideology that explains the world in toto. It demeans, ignores, and seeks to supersede other modes of knowledge when it is incapable of answering many questions or ministering to real-world concerns. In summary, the scientific worldview provides a cognitive framework that impedes full appreciation of the global environmental catastrophe.

The Probabilistic Universe versus the Kingdom of Fate

Scientists seek patterns and relations, rather than being content with observing random and chance occurrences. The paradigmatic formulation of this is Newtonian mechanics: a ball that strikes another one operates under the equation $F = ma$. Not only is the relationship vivid and tangible—one mathematical formula that everyone is comfortable with—it also operates in an exact, universal, and predictable fashion. Exact because it can be expressed in the language of mathematics. Universal because it is true for all times and places. And predictable from the previous presumptions: the same result can be expected *ceteris paribus*. Chance is only a matter of marginal errors, usually derived from faulty observations.

The hegemonic hold of exactitude, certainty, and necessity can be gleaned from the belated emergence of probabilistic thinking, which is conventionally dated to July 1654, when Pascal wrote on the topic to Fermat (Gigerenzer et al. 1989, 1). Emerging from the "medieval sense of any opinion warranted by authority to a degree of assent proportioned to the evidence at hand," the concept of probability entered the intellectual lexicon (Gigerenzer et al. 1989, 7). Notwithstanding numerous forerunners, the paradoxical convergence of mathematics, that most certain of the sciences, with probability, the study of uncertainty, in the century best remembered for the Scientific Revolution, was slow to spread. The dominant understanding of probability remained mired in the realm of necessity: subjective minds that merely misunderstand objective errors (Gigerenzer et al. 1989, 11). As Laplace ([1814] 1902, 3) observed in his treatise on probability: "All events, even those which on account of their insignificance do not seem to follow the great laws of nature, are a result of it just as necessarily as the revolutions of the sun." There is no chance—just error—in this philosophy of probability. The world remains orderly and predictable: a kingdom of fate.

The spread of statistical science in the early nineteenth century retained faith in order and predictability. Precisely when Poisson was developing mathematical probability in the 1830s, statistics—the science of the state—became equated with the study of numbers (Gigerenzer et al. 1989, 38). It was part and parcel of the erosion of rationality and the incorporation of the masses; statistical thinking shifted from "the psychology of the rational individual to the sociology of the irrational masses" (Daston 1988, 187). Mobs may be unruly and unpredictable, but statistical science offered respite: micro instability and confusion, macro stability and clarity. It reined in chance and randomness. Quetelet's concept of *l'homme moyen* similarly sought to tame the ostensible irregularity and unpredictability of the masses by postulating a social atom for the new science of social physics (Gigerenzer et al. 1989, 41–44). Just as the Congress of Vienna restored order and stability to the Europe of change and chaos after the French Revolution, Quetelet and his colleagues sought to re-entrench the intellectual understanding of the world as one of regularity and necessity (Porter 1986, 107–109, Schweber 2006, 121–122).

Probability and statistics in their first two centuries, then, are a chronicle of efforts to tame the chaos of chance. It is only in the late nineteenth century—exemplary here is Peirce's philosophical embrace of chance—that probability becomes unmoored from its foundation in law, order, and necessity. Yet these disciplines of probability and statistics developed slowly; only by 1930 were their mathematical foundations secure (Porter 1986, 315–16). Counterrevolutionary currents remained powerful, from the mid-nineteenth-century physiologist Bernard to the early twentieth-century physicist Einstein, both of whom rejected the role of chance in nature. Bernard's commitment to causal determinism was predicated on his belief about the nature of exact science. As he wrote in 1865, "If based on statistics, medicine can never be anything but a conjectural science; only by basing itself on experimental determinism can it become a true science, i.e., a sure science" (Bernard [1927] 1957, 139). Countering Bohr's interpretation of quantum mechanics, Einstein wrote to Born in 1926, "I, at any rate, am convinced that [God] is not playing at dice" (Einstein, Born, and Born 1971, 91). Neither contemporary medicine nor present-day physics would be conceivable from the standpoint of determinism.

The probability revolution comes to prevail in the world of science in the course of the twentieth century. "The most decisive conceptual event of twentieth century physics has been the discovery that the world is not deterministic" (Hacking 1990, 1). From the rarefied field of high-energy physics to the profane world of gambling, the stochastic worldview has spread. In the process, it has upended deterministic certitudes and subjective intuitions and has replaced them with probabilistic calculations and statistical analyses. The shift is clear not just in the sciences—from Newtonian mechanics to quantum mechanics or from Bernard's physiology to statistically based medical diagnoses—but in everyday life. The disenchanting mindset of big data-driven analyses expunges the human touch—from baseball coaches' hunches to stockbrokers' inspirations—in decision-making (Lewis 2003, Zuckerman 2019). We cannot escape the stochastic universe of randomness and contingency, large patterns and huge trends. In the early twenty-first century, we are living in a world of numbers, probability, and risk. From shopping suggestions to hedging financial fluctuations, we are awash with algorithmic calculations that presume a world of indeterminism. But do we truly inhabit a stochastic universe in our mind?

As much as we are aware of the idea of the stochastic world and accept life's indeterminacies, we retain a pre-probabilistic mindset. Put differently, we often think like Bernard, Einstein, and the heroes of the Scientific Revolution. Ironic then that "chance, superstition, vulgarity, unreason were of one piece" in the Age of Reason (Hacking 1990, 1), when the Age of Chance deems discourses of determinism as wrong and superstitious. The message about the stochastic worldview has not reached those who continue to dwell in a universe of order and determinism. In this worldview, science dissects nature that is necessary and certain. In resisting the stochastic revolution, we misunderstand the nature of contemporary scientific thought and practice, and remain entrenched in the kingdom of fate.

The belief in causal determinism—and the correlative expunction of chance—is a common human cognitive orientation. In a classic ethnography, Evans-Pritchard (1937, 69) explained the search for causality among the Azande of Sudan: "That [an old granary] should collapse is easily intelligible, but why should it have collapsed at the particular moment when these particular people were sitting beneath it?" A self-styled rationalist would deem it as a coincidence—"these particular people" were in the wrong place at the wrong time—and would scoff at the Zandean belief in causal necessity, especially as it is believed to occur by means of witchcraft. Yet few would resist the lure of determinism when an event is personally meaningful; a loved one in an automobile accident may occasion one to rack one's brain about all the unfortunate yet inevitable steps that led to the tragedy. In our dramaturgical view of life, we often identify something, like Chekhov's gun, that *must* go off. One does not need to be a Calvinist believer in predestination to say in this instance "there but for the grace of God go I," or the more succinct Islamic "inshallah," both of which disclose the same Zandean cognitive framework.

The Zandean belief system is embedded in everyday life—one "actualized these beliefs rather than intellectualized them" (Evans-Pritchard 1937, 82)—and may invoke external agents, such as witch doctors, who may be differentially skilled (some may be charlatans) (Evans-Pritchard 1937, 183). It would surely be an ethnocentric mis-recognition to belittle the Zandean world of necessity and the potency of witch doctors, for the self-styled rationalist might engage in the same sort of thinking, going so far as to employ their own version of witch doctors. A Stanford genetic engineer explained her reliance on fortune tellers to me in language that would have satisfied any Azande: "My marriage will not be a random event. We are all fated, one way or another. There are fortune tellers who are frauds, naturally, but who can deny the power of forces beyond our control, and some gifted people who can read them?" "You can't escape your fate" (Stepanova 2021, 329) is a garden-variety sentiment around the contemporary world, whether invoked as karma or kismet in one's choice of career or spouse. Many Millennials and Zoomers (aka Gen Zs) in the Global North, in this spirit, avidly follow astrology (Jeffrey 2023). Destiny, even predestination, rules the roost of personal life. It is hard to imagine resisting, even in the early twenty-first century, the logic of Donne's "Love's Deity"—the notion of romantic love as destiny—especially when in its thrall.

The prevalence of determinism in everyday thinking resonates well with the received understanding of the Scientific Revolution. Causal determinism, not chance, is what science provides in this worldview, and with exactitude and therefore the prospect of predictability to boot. Far from dwelling in the contemporary scientific universe of contingency, the scientific worldview abides in the certain and comforting kingdom of fate.

The Intellectual and Material Mastery of Nature

The Scientific Revolution promised to make the Book of Nature legible. Bacon is surely a major beneficiary of the Matthew effect—so memorable quotations, along with the collected works of Shakespeare, have been ascribed to his quill—and two resonant ideas are encapsulated in two phrases that are not exactly his: science as an enterprise that engages in "dissection of nature" and "knowledge is power." As Whewell (1860, 128) declared: "If we must select some one philosopher as the Hero of the revolution in the scientific method, beyond all doubt Francis Bacon must occupy the place of honour." Never mind that he was keen to supersede Aristotelian and Arabic natural philosophy (Bacon [1620] 1902, 56) when all the advances in seventeenth-century natural philosophy depended on classical and medieval, especially Islamic, antecedents (Lindberg [1992] 2007, 366). Nor did Bacon do any work that can charitably be called scientific. If he didn't exist, then we would have to invent him as the prophet of modern science who sought the intellectual and material mastery of nature, in Helmholtz's ([1891] 1912, 272) felicitous but frightening phrase.

As even a non-assiduous reader of the Bible can attest, early on in Genesis (1:28) God's likeness was given "dominion over the fish of the sea, and over the fowl of the air, and over every living thing that moveth upon the earth." The text seems to declare "man to be a free agent who has the God-given power to control nature," though the power "cannot include the license to exploit nature banefully," if only because of God's ultimate sovereignty (Sarna 1989, 12–13). As much as most human communities most of the time exploited what was available around them, the idea of human dominion over nature is modern. It would be the height of hubris to claim mastery over nature, but we moderns talk seriously about it, even or especially in the face of the environmental crisis.

Human beings have always sought to intervene in the workings of nature. Ostensibly contradicting the commonsense belief in fate, there is nothing inconsistent in our fervid efforts, by hook or by crook, to affect the unfolding of events. Intellectual mastery is in this regard inextricable from material mastery. It is precisely because knowledge informs practice, and produces discernible results, that knowledge comes to be valued. The Zandean belief in witchcraft is emplaced not only in Zandean life but also in the efficacy of magic and witchcraft. Certainly, there were no better alternatives at that particular time and place.

Scientists and believers in science may scoff at manifold manifestations of irrationalism and occultism, superstition and sorcery among the Azande and other nonscientific peoples, but the scientific revolutionaries of the seventeenth century were enmeshed in them. Astrology is as good an example as any. The arch-rationalist Adorno ([1957] 1975, 17) inveighs against horoscope as "modern big time irrationality" and "secondary superstition," but the putative founder of modern astronomy, Kepler, regularly drew up his personal horoscope and worked at the vast intersection of astrology and astronomy (North [1993] 2008, 338–45). Indeed, astrology captivated the advanced intellects of the Renaissance (Kuhn 1957, 93–94). The spectacle of the full moon casts spells on self-regarding rationalists, and the impact of lunar cycles is not only on tides but also on our biorhythms and psyches. In a similar vein, Newton's obsessive biblical investigations belie the mantle of modern science and disclose what we would call a fundamentalist Christian mindset (Iliffe 2017, 8, 401). Contemporary adherents of astrology or Christian fundamentalism, then, inhabit the mental universe of seventeenth-century scientific revolutionaries more than their successors. Bacon more than dabbled in magic and alchemy, which may have informed his notion that knowledge is power (Zagorin 1998, 40–44).

Examples of other cultures (the Azande) or other times (the seventeenth century) are all well, but we moderns are wont to claim that we *know* that science and technology are distinct from, and superior to, magic and religion. Magic is said to involve occult techniques and to harness an imagined supernatural force. Religion is belief in divinity, which may also not exist. In contrast, science stands on logic and evidence: the realm of reason and the real ("science is real," as the ubiquitous poster in my neighborhood declares). Certainly, scholarly treatises readily demonstrate the chasm before and after the Scientific Revolution.

Mayr's (1982, 22) distinction captures the gist of what many accept: "A fundamental difference between religion and science, then, is that religion usually consists of a set of dogmas, often 'revealed' dogmas, to which there is no alternative nor much leeway in interpretation. In science, by contrast, there is virtually a premium on alternative explanations and a readiness to replace one theory by another." His convenient distinction is founded on a faulty empirical premise. Any serious scientific investigation depends on dogmas—if defined as axioms or unquestioned premises—and even a cursory acquaintance with theology reveals a welter of alternative and shifting interpretations. As we saw in the previous chapter, he vehemently rejected the new classification of Archaea as a biological kingdom. We should recall that theological disputes occasion wars, and scientific debates are not always about logic and evidence. Mayr, an eminent scientist, is no expert on epistemology and it may behoove us to consider a more searching study.

Malinowski's celebrated discussion of magic, science, and religion remains a *locus classicus*. Writing in the mid-twentieth century—at the height of techno-scientific enthusiasm—he seeks to rescue "primitive" and "savage" populations from the condescending, and often condemnatory, perspective of the advanced, civilized, rational, and scientific civilization. As much as he rebuffs Frazer's evolutionary and ethnocentric schema, Malinowski retains the commonsense of modern condescension. After observing that the "natives" engage in "the disinterested search for knowledge," he suggests that magic is a means to give them "the power over certain things" and it "ritualize[s] man's optimism, to enhance his faith in the victory of hope over fear" (Malinowski [1948] 1954, 35, 79, 90). The crucial distinction is that "Science is open to all ... magic is ... handed on in a hereditary or at least in very exclusive filiation. While science is based on the conception of natural forces, magic springs from the idea of a certain mystic, impersonal power, which is believed in by the most primitive peoples"(Malinowski [1948] 1954, 19).

I am not asserting epistemic equality, and definitely not identity, between religion and science, but categorical differentiation—and condemnation—may obscure more than it illuminates. Instead, we should strive for epistemic sympathy and symmetry and employ the same standards. Is science open to all? Not only is there an intergenerational reproduction of scientists in the modern West (cf. Chise, Fort, and Monfardini 2020), but we moderns also tend to regard the work of science as opaque and dealing with incomprehensible and impersonal forces. Technology as applied science surely gives us "the power over certain things," but what do most of us understand of the basic infrastructure of modern life, such as electricity, or its more advanced articulation of knowledge, such as genomics? For the silent majority that routinely depend on technological outcomes of scientific and engineering advances—from automobiles to airplanes, from microwave ovens to internet transmissions—these conveniences work like magic or witchcraft, and what can we say about them beyond what the Azande say about their communal belief? Scientific literacy is far from being well disseminated in the self-proclaimed greatest country in the world (Kennedy and Hefferon 2019, Mooney and Kirschenbaum 2009). Put polemically, then, many moderns reside in the world of magic and adhere to the religion of science and technology (Noble 1997, cf. Feyerabend [1975] 1993, 31–32). Very few reflect on or investigate the nature of technology in spite of its centrality and ubiquity in our life: no more so than a typical Azande, one suspects. As Malinowski ([1948] 1954, 17) suggests: "There are no peoples however primitive without religion and magic." I would only excise the phrase "however primitive" to turn the proposition into a robust generalization: a cultural universal. Malinowski's "superstitious turn of mind and his propensity for magical thinking," as his biographer remarks (Young 2004, 510), is but one example.

Save for exceptional times of ferment, there is always a political institution that maintains order and stability in political and social life. Similarly, there is an epistemic regime that privileges one mode of knowledge and belief over others; some of which are tolerated, others condemned. No knowledge regime is permanent. In the early decades of the first millennium, ancient Greek intellectuals castigated the new-fangled Jewish sect as "superstition," and later "Christian intellectuals were right, from their point of view, in branding Hellenic religion and philosophy with the same term" (Martin 2004, 243). The early medieval Christian Church, surely the dominant epistemic ruling force of the place and the time, encouraged magic, which had been condemned in the Roman Empire (Flint 1991, 3, 323–26). Later Christian authorities, in contrast, castigated it (Thomas [1971] 1973, 762–66). Not only was magic taken out of the world, but so too were magicians and witches (Bartlett 2008, 32–33). Witchcraft became the epistemic enemy when there was scarcely solid "evidence that witches existed, still less that they celebrated black masses or worshipped strange gods" (Briggs 1996, 6). The hegemony of the Christian Church spelled the decline of magic.

The seventeenth-century Scientific Revolution was, in retrospect, a turning point in the supersession of religious belief by scientific knowledge in the Christian West. Never mind that magicians theorized about nature and inferred technology to alter nature, and so were in principle no different from scientists (Copenhaver 2015, 424–27, cf. Walker 1958). As we have seen, while Kepler and Newton straddled terrains that are firmly divided today—magic, religion, and science, or alchemy and astrology as opposed to chemistry and astronomy—their successors by the nineteenth century were ensconced confidently in the one true terrain of science. As Huxley (1893, 65–66) exhorted in 1860: "Extinguished theologians lie about the cradle of every science as the strangled snakes beside that of Hercules: and history records that whenever science and orthodoxy have been fairly opposed, the latter has been forced to retire from the lists, bleeding and crushed if not annihilated: scotched, if not slain." The same Christianity that had condemned magic was in turn denounced as irrational and superstitious. In short, heresy or superstition is false, irrational, and dangerous not only because it poses a threat to the dominant epistemic regime but also because it is orthogonal or antagonistic to the associated political–economic power. In condemning magic, religion changed, at times taking on a scientific cast, whether abstractly in Protestant theology or concretely in Christian Science (cf. Harrison 2015, 187–88). The enormous condescension of the triumphant epistemic regime renders medieval science an oxymoron and other forms of knowledge at once inferior and dubious (Falk 2020).

A regnant epistemic regime rarely, if ever, achieves total domination. If Victorian England, especially after Darwin's theory of evolution, was a battleground of religion and science that seemed to end in the triumph of the secular intellectual order, the early twentieth century—and into the early twenty-first century—retains a healthy representation of religious believers and defenders among scientists (Bowler 2001, 362–65). Superstition and other manifestations of irrationality are omnipresent among self-styled rationalists. The Age of Reason—and at the center of the light, in postrevolutionary Paris—is in retrospect replete with strange fads and fashions that we would call pseudoscience and occultism. Mesmer achieved immortality of sorts by proposing a science and technology of animal magnetism: "Mesmerism offered a serious explanation of Nature, of her wonderful, invisible forces, and even, in some cases, of the forces governing society and politics" (Darnton 1968, vii). The guardians of the scientific establishment sought to banish deviants and heretics from their domain, but a cursory examination of popular scientific beliefs and practices would reveal that ostensibly unscientific and irrational ideas survived and thrived. And who can be sure about their falsity and irrationality, as history is replete with examples of cranks and crackpots who turned out to be right?

In any case, there is no question that modern science and technology reign as the dominant epistemic regime in the twenty-first century. The growing prestige of natural philosophy after the Scientific Revolution superseded the epistemic authority of Christianity in the West by the turn of the twentieth century. Not only did the educated population embrace the theories of Newton and, more devastating for Christianity, Darwin, but technological applications associated with the Second Industrial Revolution disseminated the blessings of modern science to the public: from the railroad and the telegraph to electricity and the radio.

Most important, scientific knowledge came to *define* knowledge (cf. Gilson 1941, 111–15). Religion and other forms of thought persist, but they dwell in the realm of belief and other categories beyond and beside (scientific) knowledge. Who speaks of knowledge in the modern world speaks of science. The rest is, if not quite silence, then quaint or irrelevant.

Nevertheless, it is unclear how much of this hegemony is due to our grasp of scientific fundamentals, much less keeping up with the latest reports in *Science* or *Nature*. The level of scientific literacy in the United States, as noted above, should give pause to the claim of modern science as one true form of knowledge that can be defended by an informed citizenry. If we follow Huxley (1894, 310), who proposed agnosticism, then "it is wrong for a man to say that he is certain of the objective truth of any proposition unless he can produce evidence which logically justifies that certainty." The majority's faith in modern science may be more akin to the faith of medieval European Catholics than we like to think. We recall those Catholics as benighted believers who recited catechisms and did not even read the scriptures—therefore functioning as intellectual slaves, in contradistinction to we moderns who are intellectually autonomous and free. We learn in school about the superiority of science—as our medieval European counterparts learned about the true faith in the Church—but the faith is nurtured because it is inextricably intertwined with everyday life. The modern conception of religion, especially in the Christian West, is a matter of belief or theology, but for most people most of the time it was and is more about a way of life: a concatenation of tradition, culture, and faith (Smith 1963, 193–95). Religion is more about feelings than reason in and of itself: "We believe without belief, beyond belief," as Stevens said (cf. Royce 1913, 428–29). Medieval Christians followed a way of life, with rules and rituals, without much interest in theological questions and disputes. Who now remembers Arianism? I am suggesting that many of us in the modern West are no different. Science serves as a cosmology or a worldview, and the magic of technology cements our faith in contemporary witches called scientists. Science-based technology is enmeshed in everyday life, but few care how it works, and fewer plumb the depth of the intellectual edifice that is said to underpin it (Winner [1986] 2020, 5–6). How many frequent flyers have wondered about the mystery of flying on an airplane (cf. Bloor 2011)? We are awash with statistics—especially perhaps in the discourse of environmental destruction—but we tend to be unsure of the particulars of statistical claims or the basic principles of statistical reasoning (Blastland and Spiegelhalter 2013, Harford 2020).

The transition from philosophy and science to worldview and ideology is commonplace. For example, from early Hegelian essays to later political–economic treatises that disclose tensions and contradictions, Marx's theoretical architecture becomes ordered and regimented, simplified and homogenized with the march of state socialism. It is as if a sumptuous banquet at the Court of Versailles transmogrified into a cup noodle: almost anyone can rise to the challenge of pouring boiling water and waiting three minutes for a complete meal, however stuffed with excess salt and strewn with chemical additives. Simplification is exaggerated when Marx's writings become a political philosophy to justify a nation-state, an ideology to legitimize the Soviet Union or North Korea. Occasionally, there are superb simplifications, such as *The ABC of Communism* (Buharin and Preobrazhensky [1920] 1922), but the authors may be liquidated rather than feted (Cohen 1973, 83–96). A similar process can be observed in religious movements from charismatic originators, such as Jesus, to steady routinization and codification by followers operating in one or another organization called the church. The received lessons of Marxism or Christianity are fed to the masses by engineers or pastoralists of the human soul. Some who consult the original words are surprised and shocked to discover the tensions and contradictions between original formulations and later codifications. The medieval Catholic Church and the North Korean state therefore proscribed reading the Holy Bible and the collected works of Marx and Engels. Violators were burned, tortured, or imprisoned.

The structure of the epistemic regime has a hierarchy of three classes: the masses who follow symbols, rituals, and rhetoric as a way of life; the cadres who engage in guarding catechisms and rituals; and the philosophers who debate fundamental principles and truths (cf. Ibn Rushd 1961, 14). In medieval Europe, philosophers debated the deepest issues about truth and faith; priests taught and reinforced received teachings; and the rest followed them. Similarly, scientists engage in pathbreaking discoveries and debates; teachers and technicians apply and popularize them; and the rest receive the watered-down teachings and, more important, miraculous applications.

In summary, science features some of the greatest achievements of the human intellect, but most people accept it on trust, as a matter of faith. Furthermore, what makes the contemporary epistemic regime of science and technology cogent is less our comprehension of the mathematical equations and experimental results and more the wonders of technological applications purportedly based on scientific discoveries. The technological sublime, not scientific comprehension, beckons belief (cf. Nye 1994). Indeed, it is science in the form of belief and magic that provides the fodder for the faithful. The intellectual and material mastery of nature is a blessing bestowed on the masses, who remain epistemically entrenched in the kingdom of fate, all the while worshiping at the altar of science and technology.

Risk and Uncertainty, Misfortune and Injustice

For the broad public, including the well-educated in wealthy countries, scientific knowledge rests largely on a bedrock of faith—the impenetrable hierography of mathematics and the mysterious powers that issue from it—and its content rests on an outdated edifice that assures scientific knowledge's access to certainty in a world of determinism.

Why should these propositions matter for our understanding of the global environmental catastrophe? The reason is that the prevailing scientific worldview has us dwelling in the kingdom of fate. We are said to bask in the bright light of a single, certain truth, when in fact science is but one dimension, however privileged, of our multifaceted sources of knowledge.

Before we proceed, let me resolve the tension between the epistemic conviction of determinism and the wobbly faith in fatalism. If the world were *fated*, then our degree of freedom would be zero. Wouldn't we then be mired in a cesspool of despair and inaction? The determinist outlook almost always allows for the existence of consciousness—the thinking, willing self—or the phenomenal reality of constant change. Hence, any plausible system of thought introduces a place for freedom and striving. There is always a crack in the perfectionist sheen; determinism is never fatalism to the reflective mind. We believe that our cunning may evoke other forces or our prayer may reach the prime mover. The major world religions demonstrate the uneasy assertion of determinism and the necessary existence of voluntarism. Hinduism, for instance, justifies the prevailing social order, but provides room for change. One's place is a consequence of the deeds of a previous lifetime; the best one can hope for is to strive for a higher place after reincarnation. In Calvinism, to take another example, the doctrine of predestination, according to Weber ([1904–5] 1930, 111–15), did not result in passive resignation but rather occasioned a paroxysm of activities as those who believed in fate paradoxically sought signs of salvation. Fate is fine in theory but not in practice. The knowledge of fate provides guidance to intervene, and transform, the predetermined future, which is the point of technology. Through knowledge and application, the laws of nature can alter reality. No one can safely wallow in the axiom of determinism, which may be true but *not for us*. The slippage is encapsulated in the hackneyed phrase, "That may be true in theory, but not in practice." There is no difference between theory and practice in theory, but there is in practice.

The faith in fate—the modern scientific articulation of nature as the realm of deterministic laws—confounds the place of risk and uncertainty in everyday life. When Gulf War II was proceeding poorly, Secretary of Defense Rumsfeld pontificated that "there are known knowns; there are things we know we know. We also know there are known unknowns; we know there are some things we do not know. But there are also unknown unknowns—the ones we don't know we don't know. And if one looks throughout the history of our country and other free countries, it is the latter category that tends to be the difficult ones" (C-Span 2002). Although it won the 2003 Foot in Mouth Award (Plain English Campaign 2003), the distinction is decisive. Whereas risk is "a quantity susceptible of measurement," uncertainty is not (Knight [1921] 1964, 19–20). Uncertainty is *au fond* immeasurable and unknown. "It is a world of change in which we live, and a world of uncertainty" (Knight [1921] 1964, 199). I would excise the phrase "in which we live," but there is no question that modernity is marked by an acknowledgment of constant and rapid change. Rumsfeld's "known unknowns"—Knight's "risk"—can be conceived and measured, which is precisely what actuaries do. "Unknown unknowns"—"uncertainty" in Knight's terminology—are definitionally incapable of anticipation and expectation, and therefore incalculable and unpredictable.

If the scientistic outlook encourages causal determinism and scientific naturalism, then the probable consequence is that we will regard fate as a product of natural laws about which human beings can do little to intervene (save by, depending on one's proclivity, a combination of magic, religion, and science). Furthermore, the natural realm remains beyond easy human control or domination. In this situation, the age-old temptation is to dismiss natural disaster as *natural*: a matter of misfortune.

The belated emergence of the social sciences stems in part from the presumed impotence of human beings in the constitution and construction of their social arrangements. People believed, whether because of gods or genes, that irrefutable but false tautology: it is what it is. Premodern societies were usually status stratified, meaning that individuals were thrown into different slots in society. As in Hinduism, as noted above, one's station in life may be the consequence of a previous life's effort, or simply part of an unquestioned natural order, as in much of medieval Europe. Modernity is marked by an epochal shift from status to equality: in principle, at least, everyone is, or should be, equal to participate in the game of life. If social inequality appears to be reproduced across generations, then that may be explained as the result of genetic lottery, social fortune, hard work, or sheer luck. Never mind that our parentage shouldn't have such a consequential impact on our life chances, ideologies of inequality mutate to make sense of gaps and gulfs in income and wealth, among other metrics of life's rewards. Although luck may be much more important than we like to believe—yet another dimension of our failure to embrace the stochastic worldview—the scientific strain in the social sciences seeks more deterministic causes and variables.

In the past, moral luck was usually ascribed to forces beyond human control: divine, supernatural, and at times natural forces. Misfortune, similarly, was beyond human knowledge and control. When bad things happen, there isn't much we can do except to rue our misfortune. A signal transformation of modernity is to make room for human agency. Perhaps the tragedy of a flood—as much as heavy rain is a natural phenomenon—is in part a consequence of social inequality and political neglect. Those who suffered may be disproportionately poor (among other features) and government may have privileged the upkeep of richer neighborhoods. Natural disaster is not purely natural. Ascription of risk inequality transforms misfortune into an issue of injustice.

Poverty provides a clear window into the distinct worldviews of misfortune and injustice. In many traditional thought systems, poverty is a permanent state of humanity: "For the poor shall never cease out of the land" (Deuteronomy 15:11). There was always pity and relief for the poor, but systematic government redress is primarily a post–French Revolution phenomenon (Stedman Jones 2004, 199–201). Although Jesus (Matthew 26:11), among others, seems to point to both the social origins of poverty and possible political remedies, poverty has transformed from a question of fate into one of justice in the past two centuries. From civil and political rights—free speech or political participation—we take for granted that there are *social* rights, such as the right to basic human needs (Marshall 1950, 46–47). "Pity would be no more / If we did not make somebody poor" is merely one articulation of the potential for a social and political transformation that would eradicate, or at least alleviate, poverty. To be sure, some continue to believe that poverty is a quasi-natural state, whether as a form of divine retribution or natural occurrence, but our conventional understanding is that we can ameliorate income and wealth inequality, which accounts for social welfare and other public efforts to address and redress the status quo. Poverty, in this modern view, is about politics and justice.

The shift from misfortune to injustice is parallel to that from uncertainty to risk. In becoming conscious of the problems hitherto lodged in the realm of nature—and of fate—we dislodge and shift them to the sphere of society—and therefore of politics. The growing concern over *environmental justice* is part and parcel of this cognitive turn. Rather than bemoaning the unfortunate lot of those leeward or downstream, or farther away in space and time, residents of the Global South or our progeny, we incorporate them into the expansive community of care and concern, calculation and internality.

Natural disasters are coeval with human history. Our earliest extant narratives, from the *Epic of Gilgamesh* to the Book of Genesis, relate a major flood, and our past is a chronicle of volcanic eruptions, violent earthquakes, runaway wildfires, and other natural disasters or—in the language of insurance—acts of God. Major catastrophes were beyond the power of magic and witchcraft, and even when the cause seemed to reside in human wickedness, the final cause was God (and God's wrath in particular), or a force beyond human ken and action.

Modern risk, by contrast, is anthropogenic—stemming from hazards of human civilization, especially ones propelled by modern science and technology. Although environmental damage is often invisible and therefore ascribed to nature, fate, and misfortune, it is possible to shift our perspective. We can seek consequences to our actions and therefore render environmental disasters as not just natural but also social (Beck 1992, 21–22). Risk is difficult to calculate because of scientists' monopoly on knowledge claims and resources (Beck 1992, 21,29), but it is not so in principle, as lawsuits and social movements to remedy pollution damage attest. Our conceptions shift over time and culture precisely because we negotiate and struggle over their domains and significances (cf. Douglas and Wildavsky 1982, Jasanoff 1999).

Environmental decay is compounded by the asymmetrical distribution of its adverse impact. Not only do rich regions remove pollution-producing industries and processing or storing waste to poor areas, but destruction of nature—coal mining and its attendant accidents and pollutions or nuclear-power generation—is disproportionately concentrated in less powerful and privileged populations. To exacerbate the problem, many leftist critics—seduced by critical social thought that stresses class as the basic axis of inequality in modern life—long ignored the toll of environmental damage: out of sight, out of mind.

Consider the tsunami that led to the meltdown at Fukushima Daiichi Nuclear Power Plant. Was the disaster caused by something that was unknowable, dwelling in the realm of uncertainty, or was it a consequence of miscalculation, a human error? Given the propensity for earthquakes and tsunamis in the Tohoku region, it was not much of a surprise that there was a massive shake that generated a major tsunami in 2011. But geologists and oceanographers—and especially the decision-makers at the Tokyo Electric Power Company—stipulated that a tsunami that would engulf the reactors was not only very unlikely but basically impossible (NHK Merutodaun Shuzaihan 2021, 373–74, 561–63).

There is a frightening (mis)calculation at the heart of nuclear power. The entire enterprise depends on the existence of fissionable materials that may only be available for several hundred years, but we know that a longer span of time would be necessary to obviate the dangers of radioactive waste (Yamamoto 2015, 280–83). The scale of time is almost geological, and beyond the known longevity of any means of human communication. We don't have a language to communicate the existence of extremely dangerous deposits to our descendants.

Faith in science is predicated on the belief that there will be solutions, whether in extinguishing radioactivity or in cultivating a new source of energy production. There are unknown knowns, such as from scientific and technological advances, that may render otiose whatever misgivings people may have, but the belief in these unknowns is based on and enveloped by scientism, or the faith in science and technology. Never mind that many of them, such as carbon-capturing technology, have been pioneered by plants and other organisms incapable of achieving scientific knowledge.

The old Stoic wisdom that we should distinguish what we can change from what we cannot misses the shifting boundary line between the two realms. The transitions from uncertainty to risk and from misfortune to injustice expand the realm of what we can do, but that space has long been dominated by scientific and technical knowledge. The epistemic hegemony of scientism threatens to obfuscate our predicaments, as the moral and social sciences that study the space of possibility remain underdeveloped and undervalued.

Hubris

Moving from the kingdom of fate to that of risk—from the mindset of misfortune to that of injustice—and enlarging our circle of concerns are *not* matters of science and technology but of social and human sciences and other modes of thought, including spiritual, religious, and philosophical traditions. We need to supersede the intellectual monopoly of modern science, especially in its scientist articulation, to create a more pluralistic epistemic world. Science and technology have done many wonderful things, but we should not forget that environmental destruction is one of the negative consequences. The problem that science created cannot be solved by science alone—at least, it seems highly unlikely that a magic bullet or two would vanquish the problem.

A scientific worldview remains the dominant epistemic system in the world. It encapsulates the most reliable and rigorous form of knowledge and generates a powerful and efficacious panoply of tools. Its vaunted ambition and technological overreach have caused myriad dysfunctions. In brief, we must tame its overweening arrogance and dominance.

Consider, first, human dominance over nonhuman nature. By closing our eyes to the evisceration of forests or covering our ears to the wailings of animals on the verge of extinction, we persist in dominating and exploiting nonhuman objects and beings. One of the curious puzzles of humanity is that we seem to be incapable of justifying dominance by sheer power and seek to legitimate its exercise. We chop trees and kill animals not only because we *can* but because we *may*. Beyond religious and traditional foundations, modern human beings often resort to one or another justification provided by science, such as our superiority born of intelligence. We are special, perhaps unique, because we are smarter than trees or animals. Modern European intellectual history, in this line of reckoning, can be read as a series of rationalizations for our centrality and superiority. We are the sole beings that can communicate with one another, use tools, or enjoy games. Each of these propositions has been decisively refuted: trees communicate, birds employ tools, and dolphins play games (Wohlleben 2015, Ackerman 2016, Mann 2017). Animal mind and animal intelligence are no longer oxymorons but thriving areas of scientific research (Griffin [1992] 2001, Godfrey-Smith 2020). Our smug invocation of our superior nature may have puffed our sense of dignity and superiority, but its ultimate test may very well be our ability to co-exist with fellow creatures and nature.

The more we know, the more we are impressed by the complexity of the arboreal world or primate intelligence and sociality. The principle of epistemic empathy generates new and renewed understanding of the subtlety and dignity of other entities with which we live. Respect for nature, twinged by fear, is of course a garden-variety feature of world belief systems, but scientific advances have come around to the same standpoint. As Lévi-Strauss ([1962] 1985, 152) argued for the epistemic legitimacy of nonmodern, nonindustrial, and nonscientific cognitive frameworks and orientations, we should strive to achieve the same sort of respect for those of other forms of life. There is no compelling reason to exclude nonhuman beings and nature at large from our care and concern, no reason not to incorporate them into our world of justice (Korsgaard 2018, Nussbaum 2023). We should not ignore the impact of extrascientific ideas and movements, such as ecological philosophy and animal-rights movements or the directions of scientific research. It is precisely when human ascendancy over nature was manifest—for example in early modern England (Thomas 1983, 302–3)—that there was a counter-sentiment against the exploitation and oppression of animals and nature. Similarly, it is not the autonomy or autotelic movement of scientific knowledge that has led us to valorize fellow organisms or the natural environment.

The intellectual and material mastery of nature trumpeted by scientism basks in the power and the glory, but it is far from being omniscient or omnipotent. Perhaps the leading expert on infectious diseases in the middle of the previous century declared "the virtual elimination of the infectious diseases as a significant factor in social life" (Burnet [1953] 1962, 1, see also Burnet and White [1953] 1972, 263). The swagger is breathtaking. Yet the same spirit of scientistic hubris often besmirches the social sciences with their systematic misrecognition of the present and the future (cf. Tetlock 2005).

With the construction of nuclear weapons, some scientists came to recognize that their impressive achievement could spell human extinction: "I am become death," in the chief architect of the atomic bomb J Robert Oppenheimer's invocation of the *Bhagavad Gita*. Humanity seemed to suffer from apocalyptic blindness: our unwillingness to see the potential of our extinction (Anders 1956, cf. Jaspers 1958). In concert with the concerned public, the scientists who developed the atomic bomb generated a global movement against nuclear proliferation and, with luck, managed to sustain a world without nuclear-war-related deaths for over seventy-five years (Schell 2020, Freeman 2023). To be sure, there were close brushes with a potential Armageddon (Sherwin 2020), but we know that hubris can be tamed and humility can be cultivated. It is crucial to remember, however, that there was nothing intrinsic to scientific knowledge that opened the scientists' eyes and moved them to control their creation. Oppenheimer's moral cultivation and imagination, for instance, resulted from his humanistic schooling, extrascientific reading, and non-scientist friends (Schweber 2000).

It would be another form of arrogance to believe that science is the only and royal road to solving the potential global environmental catastrophe. If we go beyond the risk calculation of scientists, engineers, and executives of the Fukushima nuclear disaster, then we find that there was a wooden plank erected up the hill from the ground on which the nuclear reactors were constructed (Yamauchi 2012). The following warning was on the plank: "No building should be built below this." In recent history—literacy is a blink in the eyes of geological time—there had been a gigantic tsunami equal in magnitude to the recent "impossible" one. The risk was *known*, albeit a fact beyond the measurement and calculation of modern scientists and engineers. Although working with inadequate data and dubious models, those in thrall of scientism refused to acknowledge their limitations and therefore refrained from suspending judgment and action (Levi 1980, 441–44). As Darwin ([1859] 1869, 84) confided: "So profound is our ignorance, and so high our presumption," we should strive toward intellectual humility, and in this it is not so much science but scientism that endangers our livelihood. And what better source do we have to learn about hubris than Sophocles's *Oedipus Rex*, a product of a prescientific civilization that is now entombed in the residually respected but largely derided field called the humanities.

Conclusion

Beyond the belated nature of scientific knowledge, the scientific worldview presents an overconfident belief system. Rather than presenting a probabilistic view of the world, such as the likely and not definite efficacy of a vaccine, this outlook presents science and technology as an exact, certain, and deterministic form of knowledge that reliably provides solutions to the sundry challenges to our existence. We need to shift our gaze from fate and certainty to risk and probability. We do not and cannot know the future, but we must strive to decipher uncertainty and move it to the realm of risk. In doing so, we acknowledge the limitations of science and technology and recognize the relevance and efficacy of other forms of thought and practice. The global environmental catastrophe is unlikely to be vanquished by hubristic scientism and the deus ex machina of techno-scientific magic.

3

UTOPIAN IMAGINATION

Extraordinary news beggars belief. Karski participated in the Polish Underground and was a rare eyewitness to the Warsaw Ghetto and the Bełżec death camp. Conveying a message of the systematic massacres of Jews and others in Poland and elsewhere, he faced disbelief in the United States in 1943. The Supreme Court Justice Frankfurter initially uttered: "I can't believe you." After being challenged by the Polish ambassador who was the intermediary, the skeptical judge replied: "I didn't say I don't believe him; I said I *cannot* believe him" (Brzezinski 2013, 402; cf. Lanzmann 1978). Dwelling in the realm between knowledge and belief, Frankfurter expressed a sentiment not unlike that of Horatio: "So I have heard, and do in part believe it."

The Shoah is a common historical reference point in the Global North, retrospectively recognized as one of the gravest wrongs of human history. At the time, the recognition of radical evil diffused slowly, though not for lack of witnesses. In 1942 Fry had already written about the Jewish massacre in *The New Republic*, then a well-read periodical to which Frankfurter was a contributor. As Fry (1942, 819) concluded his reportage: "This is a challenge which we cannot, must not, ignore." But ignore is what the United States governing elite did. Winning World War II was a paramount concern, of course. There was also a strain of genteel antisemitism, which in turn may have made Jews circumspect in speaking up for their co-ethnics in the administration and the country at large. Furthermore, as the journalist Downs observed about Americans' unwillingness to acknowledge the Babi Yar massacre: "It seems that … the average American cannot accept the fact that any group of people can coolly sit down and decide to torture thousands of people … . This refusal to believe … is probably the greatest weapon the Nazis have." And there were surely other factors and forces at work. What chance did Karski and Fry have, if they could

DOI: 10.4324/9781003539407-3

not convince even Frankfurter, an intelligent, informed, and concerned Jew?

Two points stand out. First, there were so many things, large and small, that the United States—and of course the rest of the world— could have done that would have alleviated, if not averted, the Shoah (Wyman 1984, 331–40). In large part the word did not spread: "The evil in this world almost always comes from ignorance, and good will can do as much damage as wickedness if it's not well informed" (Camus [1947] 2021, 119). Second, we face the challenge of converting *knowledge* into *belief.* How does one transform information—even if it seems beyond reproach, authoritative and legitimate—into something that one really knows, with the head and the heart, or conviction?

In this chapter, I consider, once again, knowledge, though this time as something that is not abstract and ethereal but concrete and real. After noting the inevitable gap—another source of delay—between the two, I examine our pervasive penchant for inaction. Far from exercising a calculating, algorithmic mind that executes constant choice and decision-making, human beings are habitual creatures who rarely engage in considered, reflective action. Here, again, scientism—in the form of the scientistic social sciences—misleads us. Nevertheless, we do, from time to time, transform our ingrained habits of thought and action. The source, or the inspiration, is unlikely to come from scientific knowledge alone, however.

Knowledge and Belief

Knowledge comprises myriad distinct elements. Science provides the most prestigious and reliable form of knowledge, but as I argued in the previous chapter, many people have a superficial understanding and antiquated reckoning of what that knowledge entails and signifies. More important, however, is the sort of knowledge people take seriously—as something of significance, not as a piece of superficial information or irrelevant abstraction. How do we transform abstract ideas into concrete truths? There are at least three hurdles to consider.

First, recall Ryle's celebrated distinction between "knowing how" and "knowing that." The former is practical: "how to do things of a certain sort," such as how to tell a joke or cook an omelet, and the subject's "knowledge is actualized or exercised in what he does" (Ryle 1945–46, 8). In contrast, the latter is theoretical, or knowledge about rules or things that one learns in school. The thrust of Ryle's argument is that many people mistake rote retention and regurgitation of abstract ideas for knowledge proper. He contrasts "the museum-possession and the workshop-possession of knowledge" and observes: "A silly person can be stocked with information, yet never know how to answer particular questions" (Ryle 1945–46, 16). A grind may parrot Einstein's formula $E = mc^2$, but he has learned neither "knowing how" nor "knowing that." The stored information may be recalled on demand, but it remains useless, and therefore it is probably wrong to call it knowledge. One must be able to assess information and work with and *through* it.

Everyday life is steeped in practical knowledge. Most people achieve mastery over "knowing how" in distinct domains, but they are often inept at converting that kind of knowledge into "knowing that." We may be good at what we do, but we are usually not good at talking about what we do. Anyone who has sought to transmit "knowing how," for example, by teaching someone else how to drive a car, can attest to the difficulty of translating this knowledge into "knowing that," or even demonstrating "knowing how." How does one negotiate a curve by steering, accelerating or decelerating, and so on? That sort of embodied, tacit knowledge, which can be demonstrated or taught, faces fresh challenges in being rendered as disembodied, explicit knowledge that can be passed on to students in lectures and textbooks. Almost everyone learns to drive, but few can convert implicit and embodied knowledge into something explicit and disembodied.

Scientific reports—and books and articles that propound them, although intended to do so in an accessible manner—are replete with academic apparatuses (the review of the literature, for instance) that are otiose for an interested but nonspecialist reader. Neither are equations, graphs, and tables likely to enlighten the uninitiated. Statistics—the vernacular of choice in environmentalist rhetoric—merely recall a quip usually attributed to Stalin: "A single death is a tragedy; a million deaths is a statistic." Reared on an impoverished diet of memorized figures and formulae, as argued in the previous chapter, even some of the best-educated people lack rudimentary "knowing that," much less "knowing how," of contemporary science. Consequently, should it be surprising that repeating the "knowing that" rhetoric has little impact, and is unlikely to convert or convince anyone? The resonance of knowledge claims resounds in distinct registers even when representing the same source, as in a bird song and a Messiaen composition (or, horribile dictu, the impoverished anglophone "chirp" or "twitter").

As a matter of classroom, or academic, knowledge, scientific education is impoverished, which in turn is bemoaned, everywhere. The dissemination of "knowing that," such as that entailed in the impending global environmental catastrophe, has, therefore, little traction in the wider population, and the situation is dire for scientific "knowing how."

The intellectualist misunderstanding of knowledge is commonplace, and surely dominates the popular discourse on the environmental crisis. There is a long-standing Western proclivity—discernible in the ancient Greeks, if not earlier—to contrast two distinct modes of human life: the active and the contemplative. Given that the task of transmission down the generations is done by those on one side of the divide, *vita contemplativa* has consistently been deemed superior to *vita activa* (Louth 1981, 203, cf. Butler [1922] 1926, 201, 221–23). The corresponding valorization of reading, thinking, and theorizing has encouraged intellectual involution: the temptation toward hermeticism and the tendency toward longueurs. No wonder that academic knowledge often comes across as abracadabra.

The second intellectualist misunderstanding concerns knowledge acquisition. A common pedagogical model envisions a metaphorical switch in human cognition. With adequate inputs of information and formulae, the switch flips, and the mind is turned on, tuned in. In a humble imitation of the Judeo-Christian creator, *fiat lux* is enacted everywhere in the modern world, and its illocutionary force can be seen in our way of thinking of the mind being turned on and off. The light-switch model elides the mind, the central processing unit, and bypasses the complex process by which we come to comprehension. When a pupil understands a branch of knowledge, she masters a panoply of data and principles that sustain it. The mere memorization and regurgitation of data or principles hardly qualifies as mastery or knowledge. The process is often mysterious—for the pupil and the teacher—and the learned employs various metaphors to describe it: a journey through a dark wood, perhaps, in which one is lost and found. But the most compelling description remains that of the dawn: "When we first begin to *believe* anything, what we believe is not a single proposition, it is a whole system of propositions. (Light dawns gradually over the whole)" (Wittgenstein 1969, 21e).

The philosopher alerts us to the way in which we gradually grasp the landscape, from the dawn's early light to the twilight's last gleaming. Once visualized in the glory of the midday sunshine, we don't need the rocket's red glare to illuminate the contours of the landscape or to prove its continuous existence. It is not by a tree or a house that we come to learn and believe, but by the interconnected entirety that we see and therefore come to know. We believe what we see, but what we see is based in part on what we believe. That is, the sort of knowledge that we value is not a catalog of discrete data and autonomous principles but rather a "web of belief" that connects information and ideas (Quine and Ullian 1970).

If the philosopher's intuition is correct, then who or what will provide that "whole system of propositions": a synthesis, a landscape of the whole? Suppose one adheres to the classic model of the seventeenth-century Scientific Revolution and its presumed predicates on the role of human beings and nature. One believes simultaneously in the progressive and cumulative character of scientific knowledge and its role as a foundation for technical applications. In keeping with an idea inherited from the Judeo-Christian tradition that nature is "man's dominion," Earth exists to be exploited at our wont and for our want. If the scientistic, or scientific–technical, worldview is what one learned in school—reinforced and reproduced by family, friends, and media—then how is one to acquire an alternative worldview?

The most common path is to seek another synthesis, envision another landscape, which is tantamount to *conversion*. Typically, everyone embraces, in varying levels of comprehensiveness and intensity, the culture and worldview received via parents and friends or in classrooms, marketplaces, or one or another form of media. A puzzle arises when we encounter something different. Many who are forced to abjure received tradition because of force may pretend to accept the new by occluding the old, forbidden beliefs and practices, such as Marranos (Jews pretending to be Christians) or Kakure Kirishitan (Christians masquerading as Buddhists). But what of changes that are more or less voluntary? The light-switch metaphor dominates our view of how one shifts from one belief system to another. Conversion is a form of epiphany: a decisive encounter with a charismatic person or a searing speech that turns one's world upside down, or at least tilts it sideways.

Lightning rarely strikes an individual or a community, however. Our entrenched habit of mind, especially in maturity and certainly in desuetude, is hostile to something new, and therefore conversion is rarely radical. People espouse a belief system in which "the old was not obsolete and the new was not incomprehensible" (Nock 1933, 253). Furthermore, the process is far from being instantaneous and partakes of distinct steps slowly ascended or descended (Nock 1933, 212–15). We frequently experience déjà vu even when we are seeing something or someone new. The new comes to the prepared; radically incommensurable ideas and practices face faint prospects for success. Strait is the gate that admits the new, which must resonate with the facts and ideas of the extant web of belief that transmogrify over time into something distinct. Call it family resemblance: without a hint of kinship, the radically distinct faces rejection or, more likely, incomprehension.

Conversion, if we consider the testimony of primitive Christianity, for instance in Acts 28, rests on *prophecy* that resonates with the received idea and on *magic*, an impressive but credible demonstration. The classic combination of knowledge-power works wonders. But there is more: faith must resonate in the person and the community. Our belief depends on what we think is true, but also on what kind of person we think we are, and the values and meanings that we share with others to orient and situate ourselves.

The process of conversion is complicated because it is difficult to reconstruct the contingent steps that occasioned the transformation: When does the night really begin? A gestalt emerges, a paradigm shift occurs: we know it when we experience it, but it is almost impossible to know when it happened and what occasioned it: the background condition or the dramatic event? The seemingly insignificant pebble hurled at the mighty wall may result in its destruction. Any teacher will have experienced the Sisyphean situation in which a heavy rock that had been pushed up by day falls by night: a student remains mired in ignorance and confusion. From time to time, however, the stone sticks: a student learns to ride a bicycle, blows the trumpet, or solves an algebraic equation. It is hard to say what made it work, but a minor miracle happens often enough. The ostensibly immovable fortress of old status quo cracks and crumbles, and gives way to the new, however informed and shaped by the old.

Finally, as the invocation of the religious vocabulary of *belief* and *conversion* suggests, the hard and fast distinction between knowledge and belief is misconceived. In the scientistic mindset, knowledge is for science, belief is for religion, and the epistemic superiority of the former is axiomatic. Indeed, it is not clear at all whether belief should count as a form of knowledge in the scientistic worldview.

In our ordinary language usage, we say we *believe* it when we really know something—know it with a high degree of certainty. Far from its usual inferior perch, belief is at times elevated to the highest pedestal. This is the sense in which, among other highlights of high-school world history, Martin Luther says under duress: "Here I stand, I can do no other." Moved by this ethic of conviction, Weber ([1917] 2004, 92) comments in his 1919 lecture: "That is authentically human and cannot fail to move us. For this is a situation that *may* befall *any* of us at some point, if we are not inwardly dead." It is "moving" because we are not "inwardly dead"; conviction is born of heartfelt belief, not from some superficial knowledge. The head and the heart—to invoke a well-worn trope—must synchronize to attain knowledge-belief: something that Frankfurter was unable to do.

It is not only that belief may be epistemologically superior to knowledge, but also that the former precedes, and is a precondition of, the latter. Given that even arithmetical truth cannot be defined by arithmetic—the crux of Tarski's ([1956] 1983, 265–67) undefinability theorem—we circle back to the unprovability of axioms without which we cannot proceed in our thinking and investigation. Logical coherence and empirical correspondence are important criteria for truth, but we ultimately rely on the illative sense, or cogency founded on totality (Newman [1870] 1978, 280–81). An ostensibly logical incoherence in one place or an occasional empirical inconsistency does not tear down the cumulative tradition of knowledge among Christians or physicists. The web of belief can absorb a shock here and a collapse there. To take up an indisputable scientific discipline, some of the most brilliant physicists working in the early twenty-first century subscribe to string theory, which postulates a series of strings in ten (some say more) dimensions as the ultimate foundation of the universe (Witten 1995, 86). Other equally lauded quantum theorists embrace the idea of entanglement, in which two quanta share a mutually influenced state over an implausibly long distance, thereby violating cherished assumptions about space and time (Bell 1964, 199, Page and Wootters 1983, 2890–91). And some are happy to profess beliefs in both Christianity and physics, however contradictory the two systems of belief may appear to others (Polkinghorne 2007, 1). Most people—including those who are intellectually smug by dint of their graduation from a prestigious university—would find ontological statements about the world from string theorists or quantum physicists absurd, and all the more so if someone should combine them with a sincere profession of Christian faith. Yet the credentialed community of physicists has, despite squabbles, generated these truths, and untrained, ignorant outsiders have no bases to contest them. But they remain, as with any knowledge claim, squarely based on accumulated achievements that have their foundations in shared axioms and intellectual developments: convention, tradition, and community. From a strictly logical point of view, they have their feet firmly planted in mid-air, but firmly planted they feel, and without it science cannot proceed.

To assert the priority of belief over truth is not a license for irrationality, but an admission of the limits of rationality. In spite of the prestige of scientific knowledge, it reigns supreme only within the ambit of expertise, and efforts at intellectual imperialism—like all forms of imperialism—are a manifestation of overreach. The monotheistic impulse of Christianity and the Scientific Revolution, however much each depends on unprovable assumptions, strengthens the claim of the one and only truth and the consequent tendency toward all or nothing. Epistemic monopoly encourages self-surrender and castigates other competing methods and communities as inevitably wrong and irrational. Scientific knowledge is ultimately a profession of faith, however, and we should be mindful of its provisional character and limitations. We cannot live without it, but we cannot live by it alone. Or, as Donne (1957, 359) articulated it more eloquently in his 1621 sermon: "*Knowledge* cannot save us, but we cannot, be saved without Knowledge."

The ultimate test of knowledge is whether we really *believe* it and whether we pay serious attention to it and act on it. A source of action is not indifferent information—it is a recipe for torpor, like so many feeds on social media—or abstract knowledge that can only lead to velleity: an impotent inclination that does not lead to concrete action. Conviction or faith—to invoke the language that we usually associate with religion—is the sort of knowledge combined with belief that the environmental catastrophe calls for, and one that may lead to meaningful attentiveness and activity. How do we go from weak volition to a strong one: the sort of knowledge-belief that will spring people into action? Before I answer this question, we need to clarify what action is and the place it occupies in our lives.

The Antinomy of Action and Structure

The Scientific Revolution was slow to reach the study of human behavior, but the power of the scientific impulse pervades the modern social and behavioral sciences. Scientism has introduced a bevy of provocative theories and useful methods. Yet scientific overreach has systematically misrecognized the nonscientific and nonrational realms of human thought and behavior. In many branches of the social sciences, human beings are considered to engage in constant, conscious *action*. A particularly strong strain based in economics and usually called rational choice theory regards human beings as undertaking instrumentally rational choice: we are a calculating animal, weighing benefits and costs, pros and cons. Although we engage in calculations, are conscious of what we do, and even make explicit choices from time to time, we spend an inordinate amount of time following settled routines, without much calculating or choosing. In contradistinction to rational choice theory, other social scientists seek to emulate the deterministic thrust of the natural sciences and propose a structuralist account. In this line of reasoning, forces beyond our ken—social forces, structures, systems—shape or determine our behavior. Neither is tenable. Instead, we are creatures of habit: habits of thought and of behavior.

Weber's ([1921–22] 2019, 78–111) typology of action remains a *locus classicus* of theorizing about action. He draws a fundamental distinction between action, which is meaningful, and behavior, which is not. Behavior comprises a panoply of unintended, unmeaningful, and unconscious activities. Reflexes (blinking in response to a stimulus), Pavlovian responses (salivation in the face of food), and other natural or instinctive movements don't presuppose the intervention of a conscious, reflecting, and thinking subject. In contrast to behavior, action is subjectively meaningful. It is intentional and conscious: social, not natural. Weber ([1921–22] 2019, 99–102) implies that much of what human beings do belongs to the domain of behavior rather than that of action.

Weber's fourfold typology of action comprises the traditional, emotional, substantively rational, and instrumentally rational. *Tradition* is action born of emulation and reproduction. Insofar as there's not much reflection or thought invested in everyday life with its patterned rituals, traditional action comes perilously close to crossing the divide into behavior. However, there is no question that it is *not* natural, but rather social, and in principle, people can generate a discourse of intention and meaning to make sense of traditional action. Especially when challenged by new forces, tradition becomes conscious and reflective and approaches the realm of substantively rational action.

Emotion also straddles the divide; anger and sadness, for instance, seem to well up as something akin to reflex, without conscious control or subjective intention. According to Weber, however, emotional action is social. Emotion is embedded in a cultural context, which is often opaque to outsiders. Tactically or strategically deployed, manipulated emotion does much to affect oneself and others. Emotional labor, in any case, is ubiquitous (Hochschild 1983). Tradition and emotion constitute a large proportion of what we do.

Paradoxically, then, the two types of rational action are relatively rare. *Substantively rational action* engenders efforts to promote a value. A committed nationalist, for example, may sacrifice herself for the sake of her nation. Martyrdom is puzzling for outsiders, especially those unmoved by nationalist rhetoric, but her action is sensible given her commitment. By contrast, *instrumentally rational action* involves emphasizing efficient means to realize a goal. In equating efficiency and rationality, instrumental rationality is rationality *tout court* for neoclassical economists and business-minded people.

In propounding the fourfold typology, Weber is proposing an ideal type: a model. A mixture of motives or types of action may inhere in many people's activities. More important, however, the domain of human action is much narrower even than in his restricted framework. The hypertrophy of capitalist organizations renders factories and bureaucracies as domains of machine-like, preprogrammed behavior (cf. Collins and Kusch 1998, 36–38). Many workers, as cogs in the wheel, exercise limited freedom and therefore, at least by Weber's definition, are not engaged in action proper (cf. Graeber 2018, 63–65). Action—defined as subjectively meaningful, conscious, intentional, and deliberate choice and enactment—is relatively rare.

A similar point can be made about attention. In the age of the smartphone, everyone appears distracted and regrets their inattention, as if Attention Deficit Hyperactivity Disorder (better known by its acronym ADHD) has become the new normative condition (e.g., Mark 2023). However, our capacity to focus has always been suspect: the grasp of consciousness is elusive and fleeting. In the flux of life, consciousness is evanescent. Indeed, we are rarely conscious or cognizant of our attentive state, and the sense in which we speak of grasping consciousness is misleading or impossible. We merely observe attention in others or, more accurately, their inattention (Cohen 2013, 205–8). We cannot understand medieval Christian monks or Buddhist meditation practices without understanding the sheer difficulty of focus and the mindless moments that constitute time and therefore our life (cf. Kreiner 2023, Nhat Hanh 1996). Just as there is less action in the world than we think, there is less attention than we wish.

Many social scientists operate under the influence of scientism and focus on instrumental rationality to erect an explanatory edifice. The exemplary social science is neoclassical economics (or rational choice theory) with its faith in the pervasiveness of choice and its reliance on mathematization that is at once logically rigorous and capable of exactitude and prediction. As much as it may illuminate and explain artificial situations—such as the prisoner's dilemma (cf. Northcott and Alexandrova 2015, 71–78)—the empirical reality is that most people, most of the time, do not act as a calculating machine. Recognition of this fact has generated a behavioral revolution in economics, while the gulf between theoretical model and empirical reality remains vast (Thaler 2016).

In contrast, some social scientists stress *structure* and the salience of constraints and determinism in making sense of human behavior. The dull compulsions of everyday life—patterned, ordered, and ritualized activities that have been received prearranged and therefore reproduced—can be understood as obstacles, forces, structures, systems, or rules that govern human life (e.g., Parsons 1951, Luhmann 1987). This approach is no less scientistic in seeking to formulate social science equivalents of laws of nature—laws of social gravity or social physics.

The debate between action and structure remains one of the fundamental questions in social theory. As a permutation of the classic philosophical question of freedom and determinism, this debate plays out the relative significance of agency (action or freedom) vis-à-vis structure (constraint and determinism). How is it possible to resolve the antinomy of action and structure and make sense of the active and passive moments of human action and behavior? Weber's notion of action that entails subjective meaning seems irrefutable—what does it mean to act without agency, meaning, or freedom?—and, at the same time, the reality of obdurate patterns and enduring structures in our lives seems compelling. It is probably not an accident that two influential proposals at resolution appeared in the same year, replete with neologisms—"habitus" and "structuration"—but, at the risk of minimizing their considerable differences, with the same desire to valorize subjective meaning while stressing structural reproduction and therefore explaining (away) the contradiction (Bourdieu 1979, 180–82, Giddens 1979, 69–73).

Bourdieu's proposal is exemplary. His early achievement was to criticize Lévi-Strauss's structural analysis, which highlighted the priority and salience of rules (i.e., structures) that govern behavior, by underscoring the omnipresence of subjectively meaningful and strategic action (Bourdieu 1972). There are rules or structures, but individuals not only follow them but also use them for their purposes. Gift giving, for instance, is not a mechanical working out of and following rules, but entails subjects to think about timing, amount, and other component factors. In particular, Bourdieu presents the concept of habitus— disposition or structuring structure—to stress the entwinement of the subjective and the objective, agency and structure, in the making of social action. However, in his later works—almost all impressive for their empirical fastidiousness and moral seriousness—almost all that habitus does is to reproduce the class structure (e.g., Bourdieu 1979). Individuals negotiate, strategize, and enjoin habitus, but the larger canvas remains remarkably stable.

The aporia is that everyone wishes to save the phenomena (cf. Duhem 1908). In the social sciences, one of the most important is action: our subjective sense that we are free to shape the future. Parsons (1951, 543–45) creates a colossal theoretical edifice to explain reproduction—how order is maintained—and action remains the magical rabbit that can be pulled from his theoretical hat. It almost never happens, however, and it is usually illusory in any case. Bourdieu, though a radically distinct thinker, hopes to save action, but we are not told how significant transformation occurs, and much space is expended on the robust reproduction of the status quo. As a form of theoretical apocatastasis, the received modus operandi is to insist on the potential for action while engaging in the task of explaining order and stability.

Action-centric theory has trouble explaining stability, whereas structuralist theory does a poor job of accounting for change. Regarding the environmental crisis, the stress on action should make a solution facile, which it isn't, and structuralist theory should make change difficult, but there are undeniable environmentalist efforts and movements. Needless to say, proponents of various theories will proffer a series of ellipses, but can we supersede the antinomy of action and structure?

The Middle Voice and Habit

Without denying moments of freedom and constraint, one way to resolve the tension, or the paradox, of action and structure is to recall Weber's implicit recognition about the pervasive character of behavior and, among types of action, tradition and emotion. In brief, *habit* dominates our lives.

The concept of habit rarely appears in the social sciences. This neglect stems in part from the rise of scientific psychology and the seeming irrelevance of habit to the scientistic mindset (Camic 1986; cf. Carlisle 2014). Capable of neither the painstaking mathematization possible in rational choice theory nor the abstruse elaboration that is the benchmark of structuralist theory, its humdrum quality promises little for ambitious social scientists. To the extent that it is discussed, therefore, it manifests itself as a residual category. We fail to act properly and fall into a bad habit, for instance. Worse, as the Gallic phrase "Métro, boulot, dodo" (commute, work, sleep) suggests, it is a form of entrapment, something akin to slow, silent death. Following this way of thinking, habit is reflexive behavior that has been mindlessly repeated: crystallized passivity. As Pater ([1873] 1888, 250) noted: "our failure is to form habits." Hence, it poses almost no significance to and therefore can safely be ignored by the scientistic investigators, and safely left for self-help advocates who seek to reform bad habits as an enemy of rational reflection and considered action.

Its ubiquity, in addition, renders it invisible, like the air we breathe. As the great theorist of the phenomenon put it: "The heavy veil of habit ... hides more or less the whole universe from us" (Proust [1925] 2002, 136; cf. Ravaisson 1838).

Another reason why habit has been sidelined in philosophy and the social sciences lies in our grammatical structure. The Sapir–Whorf thesis suggests that language shapes our perception and cognition of reality, and a common feature of Indo-European languages is that we employ the binary of the active and the passive voice. We either act or are acted upon; we are either subject or object, choose or are chosen. In this series of binary oppositions, the pair of freedom and determinism, or action and structure, fits right in.

Nevertheless, the neglected *middle voice* survives as an in-between state that straddles the active and the passive (Benveniste [1962] 2012, 165–68). Language, after all, does not determine thought; we can transcend the former's limitation to grasp another reality (and transform language in the process). A few examples should suffice to illuminate its character. To say that "this book sells" does not mean that it goes around peddling itself or that it is merely being sold. As a grammarian who would proscribe the passive voice in composition may interject, agency is mired in confusion. Yet sale happens somewhere between the uninhibited subjective freedom of action and the overbearing objective determinism of structure.

When I say that "I live," it is surely a combination of the active and the subjective, and of the passive and the objective. Even the most egocentric and triumphalist account of the self cannot narrate life as choice from *fons et origo*. After all, we are thrown into a household and society at a particular point in place and time—as much as one may wish, one could not have chosen one's parents or one's natal tongue—and therein lie the limits of the choice-centric view. At the other extreme, it is possible to narrate life as a series of reactions to events and thereby picture oneself as being adrift through an unchosen life. It is common enough to lament a passive life—think of Henry James's masterpieces on the unlived life—but no one can be purely passive if only by dint of one's developed reflexes, instincts, and habits.

Rather, life must be recast as a narrative in the middle voice, in which one reacts and acts, navigating life by means of habits. In other words, we do not go from dependence to independence but rather to interdependence. We don't quite float like driftwood but are more like surfers who spend an inordinate amount of time bobbing in the water and occasionally ride—and at times wipe out from—successive waves, with our ingrained habits and dispositions, and rare expressions of will and freedom.

The recalcitrant reality of human life is that there are clear limits to our action and volition. No one can, as far as I know, force themself to sleep without the use of nonvolitional means, such as sleeping pills. Yet falling asleep is not quite—not completely—a purely passive process. We also fall in love, but that experience is a complex mixture of the active and the passive, freedom and determinism. One cannot force or will oneself to sleep or fall in love, but the process is far from being completely passive. One can get oneself ready: taking a bath and relaxing or going out and meeting people. Both moments—the subjective and the objective, the active and the passive—are part and parcel of the middle voice, the language of life. We fall asleep and fall in love in the middle voice.

The middle voice dominates personal and social life. Falling faintly, faintly falling, we experience a shift from one state to another. Some may require more volition, or action, such as running, though most runners do not consciously will—act—to place one foot after another. As in breathing, we can do so consciously, but we almost never breathe consciously for an extended period. Instead, we enter a state of *flow* that is rarely explicitly conscious but that is not entirely passive either. The middle voice does not imply an equal mixture of the active and the passive—there's a spectrum bounded by instincts and reflexes on the one hand and conscious and reflective action on the other—but it rules much of our action and behavior (cf. Kokubun 2017, 293–94, Lie 2021, 152).

When we invoke the middle voice, we are talking about habit. Once the heavy veil is lifted, we become conscious that it is ominipresent. It occupies the same grammatical and social space as the middle voice: neither entirely active nor completely passive. To repeat, habit is not necessarily repetitive behavior that has routinized into an unreflective pattern—though this can happen—but it encapsulates elements of the active as both means and ends: practices and goals. In enrolling both voices, the concept captures the prevalence of tacit knowledge and the embodied nature of our action. We don't reflect on how we move our fingers when we play the piano or tap away at the laptop. The common refrain to "just do it" points to the vitality of the middle voice and the limitations of the active voice (overthinking) and of the passive voice (mechanical repetition).

There is, then, a large swath of life governed by habit. It has at least two dimensions: one that moves from unconscious to conscious, and another that goes from superficial to enduring. In conventional usage, as I suggested, habit is a residual category of unreflective behavior that may be short- or long-lived. Habits of words and deeds, or thinking and acting, rule life. They are institutionalized as norms, ethos, customs, rules, and laws—an ascending order of formalization—and are strikingly robust. Social life, as a concatenation of reified habits of thought and action that manifest themselves as *norm*, becomes something like second nature. There is social entropy at work, however. They may corrode over time, react to changing contexts, or—rarely but importantly—transform because of intention or conscious thought and action.

Aristotle's discussion of *hexis* in his *Nicomachean Ethics* is illuminating. Rather than regarding habit as unreflective and unthinking routines or repetitive behavior born of objective constraints, he seeks to save both its active and passive moments. The Aristotelian notion is therefore an antinomy of the contemporary, conventional conceptualization because he stresses its active and purposive dimension. A key dimension of *hexis* is that it is purposive or teleological. In this way of thinking, we become virtuous or courageous by dint of engaging in virtuous and courageous practice: "Like states arise from like activities"(Aristotle [c. 340 BCE] 2014, 198–99). The voluntary character of habit can be shaped by one's natural endowment and by teaching, but through repetition, we become virtuous or courageous (Aristotle [c. 340 BCE] 2014, 24–25, 46). Again, both nature and education play an important role, but we learn to make pottery or play the piano by practice, and repeated practice makes for a potter or a pianist (learning and repeating mistakes leads to a less-skilled potter or pianist). There is the subjective, voluntary moment but there is also the objective moment, which is a product of habituation, or repetitive practice. We become habituated to exercise virtue, demonstrate courage, craft a great piece of pottery, or skillfully play the piano—all telos with virtues that are born of habit: preparedness or disposition. There is also a negative face, however: addiction to which we become habituated.

What makes Aristotle's formulation distinctive is that he stresses the potential not only to pursue a particular end (telos) but also to change it. Habit may be enduring, but it is not destiny. Later theorists, in contrast, have tended to stress its stultifying character. Aquinas's elaboration of habit as *habitus*—Latin translation of the Greek *hexis*—views it as potential or power to act based on repetition. Neither voluntary will nor passive repetition, he stresses its enduring character (Thomas Aquinas [1485] 2012a, 431–81). Similarly, Bourdieu, who has done more than any thinker to revitalize the concept, highlights the entrenched quality of habitus but, in so doing, loses its teleological orientation and transformative possibility. In contrast, what differentiates the Aristotelian notion is its lability and mutability. We can embrace a new telos and thereby strive to transform an entrenched habit. It is this freedom that makes his conceptualization not only empirically plausible—we do decide to become different, to master a new craft, to forge a new path—but superior to those ideas that stress the determinist impulse that would render human beings as the objects of learning and socialization: the prisoners of past ideas and practices.

If we place habit in front and center, then we would see that the language of choice is often deceiving. Decision-making—entailing picking one possibility from a set of possibilities—dominates the way we think about human life in the modern West. The ruling imagery of the crossroads visualizes choice as privileging one path over others. The road chosen and the road not taken become the basis for our sense of responsibility and the potential tragedy of the human condition. That is, we become what we are by choosing. In the habit-centric conceptualization, we inhabit and embody a telos or a way of life to which we habituate ourselves. We follow the Nietzschean injunction for us to become what we are, with the caveat that we pursue a path that cannot be considered a choice from the conventional point of view.

It may seem strange to minimize the place of choice in our life, but we might take a leaf from other traditions of thought. Confucius, whose thinking accords with the Aristotelian notion of *hexis*, envisions human beings as habitual creatures who operate in the context of social norms, formalized as rituals and laws. When we engage in rituals of ordinary life, such as greeting, we do so effortlessly but not mechanically. Spontaneity is born of a mastered disposition. Yet we don't engage in choice because life is a path without crossroads (Fingarette [1972] 1998, 20). We undertake a journey of life—the path or the way—much as a stonemason may spend time carving and polishing a stone (Confucius [c. 5th c. BCE] 2003, 6–7). I am suggesting that there are different paths, but we are, most of the time, on one and only one path.

Bergson offers another articulation of the same perspective. Although we are wont to see life as replete with choices and decisions—and therefore with free will—in the form of crossroads and forked paths, we don't decide or choose as much as we change and shift ourselves (and our impulses, desires, tendencies, directions, and hesitations) "until the free action drops from it like an over-ripe fruit" (Bergson [1889] 2001, 174). Choice, in this style of thinking, is a retrospective construct that reifies the past and justifies the present. We reconstruct the forked path and thereby misrecognize the constant confusion and flux of the self and the world. Far from the image of calculating the pros and cons of distinct choices, we make up our mind from a complex concatenation of habits, feelings, reflections, and much more.

We do not so much choose as reorient ourselves when we change habits. Consider, for instance, therapeutic action. We seek to come to terms with our past; some features are beyond our ability to change, and others can be worked on. We can come to avow care and concern for another set of habits and take responsibility for the new self (cf. Fingarette [1969] 2000, 147–48). Beyond the Stoic insight that we should focus on what we can change—and forget those things that we can't do anything about—the reality is that we shift the boundary of what can and cannot be changed. Another example is conversion, which was discussed above. Why do we change when we are governed largely by habit and tradition? The language of choice misleads us. We don't choose as much as we come to new knowledge and identity (Fingarette [1972] 1998, 22). We look within to change, with the proviso that what is within, at least in the realm of ideas and ideals, came from without.

Choice is an intellectualist construct. We frequently forge arguments to justify what has been decided. Those who seek advice often know what they would like to do. As Pascal ([1670] 1995, 158) quipped: "The heart has its reasons, which reason itself does not know." More to the point, as Bruce Lee exhorted in *Enter the Dragon* (1973): "Don't think. Feel!" We think and feel, just as much as we need to know and believe in order to act.

In contrast to rational choice theory's focus on calculating costs and benefits or pleasure and pain and acting to maximize one's utility, or the structuralist stress on external forces, the standpoint of habit and norm presumes the salience of deploying disposition to positive activity. In a world of ignorance and sloth, there isn't much hope, but informed, concerned, and conscientious people can respond to real-world anxiety and emergencies. Above all, we can prepare, but the frisson before the *Ungeheuerlichkeit*—the terrible or the monstrous—can evoke not just anxiety but also action for the habituated. A shock of recognition may occasion a gestalt switch or a paradigm shift to a new regime of knowledge and action.

In short, we are creatures of habit and often seem static, but we can, and do, change. We may seem inert and resistant to reform or revolution, but the reality is that things change all the time. How does this happen? In part the central insight of deterministic thought—that we are thrown into a particular place and time that we had no hand in choosing or constructing—is irrefutable. But we can and do create ourselves to master one or another way of being. Whence this impulse?

Ideology, Utopia, and Perfectionism

Though it may strike some observers as an inert, unchanging reality, the world is, in fact full of action and transformation. In large part, change is constant, born of the unintended consequences of natural and social forces. All the same, will or freedom has erupted repeatedly. As much as we may be a product of genetic inheritance and received constitution, or the accumulated dead weight of tradition, parenting, and schooling, there are people who shift the path they have been treading or change the way they think and live. As a Nobel Laureate rhapsodized: "the times they are a-changin'." In point of fact, change is constant, but we also come to recognize how difficult it is to change intentionally. How do we escape the antinomy and the aporia?

The dialectic of utopia and ideology is central for us to emancipate ourselves from the stranglehold of the status quo and pursue a distinct path, however constrained by prevailing dispositions and structures (cf. Mannheim [1929] 1936, 209–10). *Utopia* comprises new ideas and visions that transcend the status quo. It is a protreptic, or what Emerson called perfectionism, that proposes an alternative vision of human life (cf. Cavell 1990). In the language of German philosophy, one can propose a counterfactual standpoint, or a philosophy of "as if," from which we view reality (Vaihinger 1911). The unlikely assertion of human equality in the face of pervasive inequalities is but one example of utopian ideas and visions. If social inequality is a mere effusion of natural inequality, then the world is as it is by nature and necessity. By proposing a counterfactual of equality, social inequality becomes a staple of politics. In the vocabulary of the previous chapter, a utopian standpoint transforms misfortune into injustice.

From the Augean stable of skepticism and determinism, where we believe nothing can change, arises hope that through action we can change the ostensibly immutable reality. It rests on the recognition that: "We could change the whole of society tomorrow if everybody would agree. The real obstacle is that every individual is bound into a system of established relationships which to a large extent hamper his will" (Mannheim [1929] 1936, 293). Needless to say, there are other constraints, including biological and infrastructural ones, but the most pernicious may be the unreflective embrace of calcified patterns of thought and action. Yet they can be changed in part via utopian ideas and visions, modulated and shaped by habituation to new norms.

Utopia constitutes a distinct standpoint from which to view the status quo and to transform it. Once a worldview (a former utopia) becomes entrenched, then its norms and habits become reified as policies and laws, and eventually penetrate every nook and cranny. To be sure, it is never total—there are older utopias and new ones—but the regnant reality may appear unalterable, a form of second nature. A simplified historical example may illuminate this point. In the antebellum US South, slavery was a way of life sanctioned by the loftiest theology and jurisprudence. The idea of racial equality was resisted at every turn, even after abolition and for a century after the Civil War, until the advent of the civil rights movement and Rev. Dr. Martin Luther King Jr.'s incantatory "I have a dream" speech. The civil rights movement embodied this utopian ideal—and though some social scientists will insist on its superstructural and epiphenomenal nature, we cannot live without it. The ideal of racial equality is not a simple product of scientific research but something that came out of the nonscientific language of Jesus, Gandhi, and others (and it is fitting for this book that Thoreau influenced both Gandhi and King). Over time new norms and habits of interaction, new rules and laws, entrench themselves. The utopian vision seeks to expunge racial inequality and oppression from every corner of our thought and life. What was once unimaginable—an African American president of the United States—becomes a reality. Compressed in this truncated version, it seems like a parable or perhaps a fairy tale, but—as argued in the previous chapter—without conceptual shifts, such as poverty being seen as social rather than natural, it is difficult to understand the massive historical transformations that dot the course of human history.

Disrobing the cloak of habit, we seek at once to fashion and don a new vestment, whether from recognizing the frayed fabric or the noxious odor of the comfortable clothes or from being bedazzled by the attraction of the new. From time to time, then, we come to believe that the dead weight of "habitualization devours works, clothes, furniture, one's life" and instead revel in the possibility that "one may recover the sensation of life" (Shklovsky [1917] 1965, 12). For Shklovsky, art is the privileged form of unfamiliarizing or defamiliarizing the received habit, and making possible the new—utopian or revolutionary—that transmogrifies an unlived life into *lived* life. The alternative is "to burn always with this hard, gemlike flame, to maintain the ecstasy" (Pater [1873] 1888, 277). Idealistic and hyperbolic as the thoughts of Shklovsky or Pater may be, few have never experienced a moment of enthusiasm when life open up with possibilities for us to savor the ecstasy of *vita nova*.

In my rendition of the dialectic of utopia and ideology, the primary agents of change are not free-floating intellectuals, as in Mannheim's pioneering account, but rather people who embrace one or another vision of environmentalism, habituate themselves into an environmental way of life, and thereby configure an environmental self. Furthermore, the primary mechanism is less intellectual and more emotional: people fall in love with a way of life—a utopian ideal and vision. As Aquinas intoned: "The desire for contemplation proceeds from the love of the object, because where love is, there is the eye" (Thomas Aquinas [1470] 1933, 1177). The epistemic principle of *ubi amor, ibi oculus*—where love is, there is the eye—is seemingly romantic and rare but in fact prosaic and common. When people fall in love with another person or a pet, somehow flaws disappear and the object of love transforms not only itself but the world around it. One embraces another perspective and beholds a new world. The process is almost always in the middle voice, which does not mean—as I argued—that people cannot prepare for or pursue the utopian idea or vision (just as people can prepare to fall asleep). Without love (and preparation and the disposition to fall in love), we cannot possibly plumb the profundity of Beethoven's late string quartets or the mystical beauty of Rothko's canvases. Love can be blind—the epistemic principle can mislead us—but without it, we would only hear loud noise and see splattered squares.

For people to fall in love with an environmentalist utopia entails a shift in how we view nature and ourselves and an endeavor to habituate ourselves to the new norm of environmentalist thinking and action. The knowledge must, as I suggested, transmogrify from that of idle curiosity to conviction: the knowledge-belief that is capable of action. Who or what grabs your attention is in this regard who or what shapes your belief, and it is not a purely intellectual process but a deeply emotional one. We might take a cue from philology and recall that the etymology of *stupiditas* is "to be stunted," whereas that of *sapientia* is "to savor" (Vaan 2008). The shift from knowledge (*scientia*) to wisdom (*sapientia*) behooves us to nonstupidify ourselves by cultivating our senses and sensibilities. The history of humanity is a history of feelings more than one of logic (cf. Nock 1972, 963). Do people stop eating animals for intellectual or scientific reasons? Some may, as others may merely follow the prevailing norm or abide by the law. But the transformation arises from a utopian idea and vision: a combination of knowledge and belief, intellect and emotion. We come to know and believe, with all our heart, that it is wrong to eat animals or to destroy nature. To repeat, we fall in love with a new vision, a new way of life, and, in doing so, repudiate the old that can only seem repugnant.

In cultivating a new mode of living—habits of environmentalism not just in theory but also in practice—we open the possibility of a world in which we may yet avert the catastrophe. Conventional behavior of the past, such as keeping lights on all night or tossing litter on highways or in parks, becomes matters of negligence, shame, and moral turpitude. Instead, we seek to institutionalize good environmental habits, such as reducing and eliminating standby mode that wastes a great deal of energy, and thereby shape infrastructures and institutions to realize the new ideal. We will then begin to rue, with a great deal of contempt and condescension, the spendthrift way of life that almost destroyed human civilization. Many citizens of the contemporary United States cannot comprehend the racially blighted past of their own country. Even self-proclaimed reactionaries resist a restitution of the racist past, much less slavery. More prosaically, almost everyone shudders at drivers who litter the freeway or ride without a seatbelt fastened when these habits and norms were common and normative a generation ago. Automobiles may have colonized the world, but we can decolonize it. We achieve a new vantage point from which to tut-tut the benighted behavior of yesteryears. Moral repugnance is a product of knowledge-belief.

Ecological or environmental thought is a utopian project. There are many articulations of environmentalism, and one may find inspiration in thinkers as diverse as Plato and the Buddha (Kaza and Kraft 2000, Lane 2011). In spite of the diversity, one common feature is the insight that we—human beings and nonhuman beings—are all part of one nature, inextricably connected. One such idea is the view of Earth as Gaia: an organism (Lovelock 1995, cf. Latour 2015). There is no more striking image than the revelatory photography from Apollo 8 that showed, for the first time, our planet in its entirety, resplendent in blue, green, and white, floating precariously against the expansive dark universe (cf. Gladstone and Lundy 2018). *Earthrise* is just one image, with shifting interpretations, but it is precisely the sort of source that catalyzes a new consciousness. Many people began to fall in love with the utopian idea of environmentalism from that image (however they may have been prepared to do so from earlier background knowledge). It is from the new epistemic vantage point that we develop new knowledge and politics (cf. Jasanoff 2010). Without something close to conversion, we cannot begin to alleviate and possibly avert the environmental catastrophe.

Reforms, revisions, and revolutions in our habits and norms of environmentalist thought and action must be institutionalized to reach and transform every nook and cranny of our past, destructive ways. They must come from above and below, byways and sideways. The empirical cast and the experimental character of environmental policies will perforce generate limitations and failures, but the point would *not* be to dictate from above nor to fear the unexpected (cf. Sabel and Victor 2022). But the effort entails also the highest reaches of national and international government, and the realization that we need to accelerate and expedite the realization of the environmentalist utopia (cf. Sharpe

2023). A humble social theorist, for instance, must begin to rethink her intellectual task; the social sciences should be reconstructed to encapsulate the new reality of environemntal crisis. The institutionalization of environmental thought and action is tantamount to a collective work of humanity, and one without which we will very likely not survive, much less thrive.

Utopian ideals, once institutionalized, may become routinized or calcified ideas and practices that justify and sustain the status quo. Ideology may in turn narrow our horizon of thought and the very possibility of transformation. Furthermore, utopian ideals and visions are inherently neither right nor left, not reactionary or revolutionary, and certainly not necessarily positive or good. Put polemically, Nazism or Stalinism was a utopian project that may yet make a comeback. There are no simple algorithms or formulae to choose among the available utopias. There is no reason to believe that there is a simple solution to our predicaments. Humanity may yet fall in love with totalitarian dictatorship and a way of life that would accelerate our exploitation of nature. Uncertainty is the price of freedom.

Proleptic Knowledge and Post Facto Intention

If utopias are possible and realistic—however beset by dystopias and ideological calcification—then we know that we are not doomed. The optimist in us may be right.

The most compelling reason to believe that we may be able to avert disaster is that some people foresee it. I had alluded to the US tradition of nature thinkers—conservationists and environmentalists—and we all know about Carson's blockbuster or Bookchin's (writing as Haber) prescient salvo against the ills of industrial pollution in the early 1960s. Yet as early as the end of World War II, concerned authors had alerted the reading public to the peril of environmental destruction. *Our Plundered Planet* is explicit in its tirade against our "conflict with nature," which may usher "ultimate disaster greater even than would follow the misuse of atomic power" (Osborn 1948, 9). In the same year Vogt (1948, 135, 189) lamented our "march of destruction" and the potential "national suicide" through environmental destruction. Whatever the limitations of their information and analysis, we cannot say that we have not been warned.

The future is *open* and we can generate vision and intention that commit us to change the world by transforming the way we live (cf. Danto 1973, 25–6). The modal mindset in the Global North is to seek the techno-scientific magic bullet and to rely on scientific, technological, and administrative expertise. As I argued, we must be mindful of the belated nature of scientific knowledge and the widespread misrecognition of its nature. Yet it is in thinking itself the only legitimate form of knowledge and the only possible source of solution that the scientistic worldview is most misleading. An alternative, utopian vision that will help us avert the environmental catastrophe is most unlikely to come from techno-utopian schemers.

A utopian ideal and vision can proffer an alternative orientation and telos that would order new habits of thought and action, norms, legislation, and laws for an ecological future. It is the power of imagination—much derided, faintly valued—that, drawing on the vast repertoire of human experience and expression, we need to cultivate.

POSTFACE

"Where is Everybody?" wondered Fermi around 1950 (Jones 1985, 3). By "everybody" he meant extraterrestrials and suggested what came to be known as the Fermi Paradox: the universe should be teeming with sentient and intelligent organisms beyond those on planet Earth, but, as far as we know, we have not come into contact with them. There are at least seventy-five resolutions of the Fermi Paradox (Webb [2002] 2015), but one disturbing thesis is that technologically advanced civilizations destroy themselves. Given that every human civilization has collapsed—from the Sumerian and Mayan to Roman and Holy Roman—the hypothesis is plausible (cf. Tainter 1988, 51). Our previous means of destruction have been limited, but a full-scale self-extermination is now within reach. Indeed, the proverbial sword of Damocles has hung over humanity at least since the invention and proliferation of nuclear weapons (cf. Thompson 1980).

Still earlier, Wells (1920, 594) famously intoned, "Human history becomes more and more a race between education and catastrophe," and all he had to go on was the devastation of World War I: poison gas, tanks, and propeller planes. It is surely cold comfort that humanity survived the even more devastating World War II and thus far has prevented a nuclear apocalypse. Although the United Nations—the most robust realization of Wells's preferred solution: a world government—emerged after the last world war, it is difficult to be complacent about the level of international cooperation, especially in light of the environmental crisis (Mazower 2012, 332–42). We, especially in the Global North, have survived and thrived, but how much longer?

The question is not whether the world will end in a bang or a whimper, but how it will weather the coming storm and muddle through the quagmire. London Bridge may be falling down—or being swallowed up by the Thames—and all the mad scribblings about the environmental catastrophe may be at once proven prophetic and rendered

DOI: 10.4324/9781003539407-4

irrelevant. The risk of sudden extinction is low compared to that of slow, searing suffering. Billions, for instance, will flee the extreme heat of their native habitats and seek refuge in colder climates. It is wonderful that grapes will grow in Greenland, and perhaps there will be time to enjoy the bouquet of their delectable wine, but millions will clamor for them and seek to land on the hitherto frigid and forbidding island. What are the likely outcomes? War over Fortress Greenland? Genocide of climate refugees? Environmental wars (cf. Welzer 2008)?

Given the increase in natural disasters from extreme weather events, the call to "do something" will turn shrill as mortality mounts. It is one thing to forgo air-conditioning in the Global North, but another thing altogether to reach wet-bulb temperature (heat death) and to face existential insecurity, such as submerged homes and food shortages. We may therefore face a situation that approaches what political philosophers presciently called the state of nature, in which—in perhaps the most famous phrase from that intellectual discipline—life would be "solitary, poore, nasty, brutish, and short" (Hobbes [1651] 1996, 89). Before we descend to the projected war of all against all, and in unbearable heat for good measure, we are likely to seek salvation in one or another form of authority to restore a semblance of order. The totalitarian temptation—the price of a dictator and a vast disciplinary apparatus may pale in significance compared to the cost of chaos and needs—becomes more attractive precisely as the powers that be promise to "do something": to act fast and furious (Lie 2007). Democracy, whether in the form of knowledge legitimation or political deliberation, tends to be slow and may seem ineffective against the antidemocratic alternative. Never mind the world government: the state of exception, of emergency, will likely validate the old principle that all politics is local, and bandits, politicians, and warlords will pursue local advantages and thereby exacerbate the calamity that requires cooperation. Early casualties of the global environmental crisis may therefore be any semblance of global commons or the survival of scientific knowledge and political democracy.

The global environmental catastrophe is not just about natural destruction, but also about everything else. If the idea of ecology or the environment emerged belatedly because of its elusive and expansive character—in short, it's about everything and how everything is connected—then its significance will emerge precisely in its extensive impact, reaching every periphery of our world. The Hobbesian state of nature or a totalitarian outcome does not exhaust dystopian possibilities. Science fiction—if we anoint Wells's *The Time Machine* (1895) as its foundational text—has been a series of speculations about the post-apocalyptic world. The enduring narrative structure of class division and human degeneration, beginning with Wells's Morlocks and Eloi, is neither edifying to contemplate nor alluring to anticipate.

Beyond a world government, Wells was keen on the potential of education. It is surely not a panacea because it has, at least in part, brought us to the present predicament. Science and technology, as I have repeatedly said, have brought blessings, but they have also caused the environmental crisis. Technology will shower largesse, but it may enthrall us and emit dysfunctions (cf. Ellul 1954). The Promethean condition may be universal, but who will educate the educators? The promise of the Enlightenment—intellectual autonomy and mastery—rings hollow in the face of calamities to come.

Notwithstanding the dystopian prospect, the resilience of well-tempered enlightenment, along with the potential of ordinary virtue and utopian ideas and visions, may yet prevent or alleviate the worst consequences.

In the aftermath of the Shoah—in the golden age of social psychology—several experiments seemed to show that conformity to peer pressure and obedience to authority are cultural universals. In particular, the Milgram experiment appeared to demonstrate that ordinary Americans were willing to inflict damage, to the point of potential death, to people when they were ordered to do so by authority figures. As Milgram (1971, 180) lamented: "I am forever astonished that ... young men who were aghast at the behavior of experimental subjects and proclaimed that they would never behave in such a way, but [in the context of war] performed without compunction actions that made shocking the victim seem pallid." Yet Gamson, Fireman, and Rytina (1982, 147) have shown that people do resist unjust authority and engage in moral action. Lessons of the Holocaust, for instance, sink in, and resolve individuals *not* to repeat an evil deed. Indeed, we know that some people, who were highly conscious of extreme risk, defied terror and engaged in the rescue of persecuted Jews (cf. Tec 2013). If it has been done, then it can be done again.

It is a commonplace misapprehension that only heroic, superhuman courage will save the world: the stuff of Hollywood blockbuster films. People can, however, enact the ordinary virtue of decency, with potentially remarkable repercussions. The Japanese diplomat Sugihara, stationed in Lithuania and elsewhere during World War II, issued visas to European Jews and saved several thousand lives. Asked about his "heroic" deeds, he was inclined to stress their nonheroic, mundane character (Furue 2020). His actions were ordinary precisely because he never placed himself in serious harm's way, though few good deeds go unpunished and he was dismissed from the ministry.

Ordinary virtue is within our reach, and in concert with education—not just information gathering and analysis, but also consciousness raising and sentimental education—we can alleviate, if not eradicate, environmental stress and distress. It is bracing to realize that a dog fighting a bear was a form of entertainment that was extremely popular when Shakespeare's plays were being premiered, and it was the norm to dump waste wantonly onto streets or into the Thames. Yet habits and norms change, and our moral psychology, too, that can no longer tolerate the violation of fellow beings or harm to the natural environment. We can become habituated to the new norm of co-existing peacefully with nature, whether in condemning cruelty to animals or refraining from uncaringly disposing of waste outside. Virtuous cycles can be expanded and intensified.

The coming catastrophe may make my proposal seem incredibly naïve and hopelessly idealistic against realist platitudes—that we live in a dog-eat-dog world. To be sure, few have seen a dog eat another dog, but self-defined, hard-headed realists confidently proclaim ideological shibboleths as irrefutable truths in a manner similar to those who are skeptical of global warming or at least its imminence or severity.

Sustained reflection has always resisted crackpot realism, however. Neither the apostle of the free market nor the arch-pessimist in European thought says what conventional wisdom has them say. Smith is often taken as an apostle of self-interest, but he presents another view. "Human nature jumps back with horror at the thought" of human callousness—someone who would privilege his little finger over "the lives of a hundred millions of his brethren"—because it is neither "the strongest impulses of self-love" nor "that feeble spark of benevolence" that dominates our reaction to the news of disaster. Rather, it is "reason, principle, conscience" or "the love of what is honourable, and noble, of the grandeur, and dignity and superiority of our own characters" that is foundational (Smith [1759] 1976, 137). So much for Smith as the prophet of the free market red in tooth and claw.

In a similar vein, we remember Schopenhauer ([1841] 2017, 177)—if we remember him at all—as an arch-reactionary and ur-pessimist, but his ultimate ethical maxim was: "*Neminem laede; imo omnes quantum potes, juva.*" First do no harm, as Hippocrates enjoined, but help others as much as one can. And the injunction applies equally to our lived environment. It is faintly embarrassing to speak of human nature—don't we know about the diversity and plasticity of human beings?—but it is a rare person who lacks ethics and empathy, compassion and conscience: impulses and thoughts to care and to take care of others (cf. Mengzi [c. 300 BCE] 2008, 45–47). As the French conflation of the English "consciousness" and "conscience" suggests, cognition is inseparable from ethics. If the past illusions of human nature—our self-ishness, our rapacity—are misleading, then we might hazard a hypothesis of our benign, sympathetic potential (cf. Sahlins 2008, 111–12). We may pillory the naturalistic fallacy (Moore 1903, 44–45), but we ascribe one or another characteristic to nature to justify our norms (cf. Daston 2019, 68–69). Whatever the genealogy of morals or the nature of human nature, we remain moral beings.

The received leftist response stresses—correctly—the sheer power of capitalism, with its unrelenting pursuit of profit—and its allied institutions in exacerbating environmental destruction. Like other major struggles, averting the global catastrophe need not wait for the end of capitalism or be sidelined to anticapitalist politics. Given the truncated time span in which we have to act, the coming global environmental catastrophe cannot wait for the secession of the current economic structure (as much as double suicide remains a distinct possibility). No less an authority than Trotsky ([1930] 2017, 234–35) declared it was Lenin's decisive individual act and unilateral decision that made the Bolshevik Revolution possible. Perhaps the enterprise was a disaster, but the crucial point is the possibility of individuals to shape and shift the world. Hard-headed scientists find it difficult to brook voluntarism, but—as I noted—we cannot extinguish the spark of freedom. Human history cannot be written without it (Zweig [1927] 1964).

Consciousness raising and personal action are inadequate in and of themselves and can be moreover misguided (e.g., Mann 2021). Electric vehicles, for instance, present themselves as part of an ecologically viable future, though the search for lithium and other rare earths for the battery wreak havoc on nature and dissipate energy (Lakshmi 2023). Recycling, to take another favored personal solution, long relied on dispatching consumed materials to China and elsewhere and were often merely stockpiled or incinerated (Katz 2019). We have a proclivity to pick the lowest-hanging fruits that do not tax us, but these may in turn be futile except as a form of virtue signaling. In so doing, we may mistake corporate greenwashing for genuine sustainable policy. Ecologically correct ideas and policies turn out to be ineffective or, worse, exacerbate the perilous situation. The proliferation of the smartphone and social media—everything from each superfluous "like" message to mining for bitcoin—places an inordinate demand on electricity generation and transmission (Walkley 2023).

As tempted as we may be to reach pessimistic conclusions, the reality is that we have come to know the abuses and reformed some of them. How were they possible? In part it is the collective work of journalists, activists, and employees of problematic firms that disseminated the news that in turn contributed to criticisms and improvements. It'd be facile to lambast attempts to improve electric vehicles (where would the electricity to power the vehicle come from?) or to establish sustainable development goals (are they sufficient to halt, much less overturn, global warming?) or to institutionalize corporate environmental responsibility. But we would be foolish to condemn them ex cathedra, as people working in multiple domains are a necessary feature of any utopian project. Recycling, for instance, should not be jettisoned but enhanced (Jørgensen 2019).

Put simply, any utopian vision, whether the techno-scientific or the environmental, is neither homogeneous nor definitive. Critical discussions must continue—here we should take a leaf from the corrigibility of scientific knowledge—and we must strive to think and rethink our actions and policies. Personal troubles and solutions cannot be severed from larger, structural problems and policies.

Beyond promoting enlightened education and ordinary virtue lies the necessity of vision. It is the telos to which one orients oneself—the new normal—and orders new habits of thought and action. As I noted in the previous chapter, we need utopian ideas, and we must cultivate our imagination. As Frye (1964, 140) wrote, "The fundamental job of the imagination in ordinary life … is to produce, out of the society we have to live in, a vision of the society we want to live in." But whence these visions and imaginations of the future?? How would it be possible to suspend our disbelief and cultivate negative capability? Like proto-environmentalist ideas that have been with us—if at times silently gathering dust in the libraries—they are in the repertory of accessible sources. It is not as if we are in want of viable visions of life that can obviate the global environmental catastrophe.

British Romantic poetry still resonates after two centuries and not merely as a redoubt of lovelorn youth. To stir the soul of environmental consciousness, there are few verses as resplendent as those in the poem we remember as "Tintern Abbey." For it (re)discovers nature and its "beauteous forms," kindling "sensations sweet, / Felt in the blood, and felt along the heart." Against "the sneers of selfish men" and the smokestacks that blight the view and the air, the poet affirms our "living soul" and "cheerful faith" in the splendor and sublimity of animals, trees, and landscape. Or, as Keats affirmed, "The Poetry of earth is never dead."

There are other traditions of thought. From the vast corpus of US environmental discourse, consider only Marsh (1864, 35): "Man has too long forgotten that the earth was given to him for usufruct alone, not for consumption, still less for profligate waste." Antiquated gen-dered noun and pronoun aside, these are not bad norms to live by, as is Jeffers's exhortation to "unhumanize our views." The environmental imagination has been, seemingly always and already, here with us to guide our path toward new norms and habits (cf. Buell 1995). After all, most cultures most of the time envisioned human beings as part of nature and nature as part of us. The current crisis may spawn a new vision, a yet-unthought-of utopian idea. The terrifying spectacle of nature gone mad—a pathetic fallacy—may provoke the unholy beast of non-enlightenment thought that leads us to a world impervious to logic, rationality, and other values that may be viewed in turn as having unleashed the disaster in the first place.

In our received understanding of the Scientific Revolution, a new way of seeing and knowing superseded revealed religion. What we need, however, is less a religion of science than a religion of nature: a shift from an anthropocentric to an ecocentric worldview in which we no longer dominate but become a (small) part of nature. That is, we pay attention and engage in actions that valorize our fellow creatures and the world that we share.

It may seem incredible that a book that draws primarily on other books—and hence ensconced in the world of the educated elite—would feign the potential for change. At the very least, the emotional revolu-tion is likely to spring from spiritual or visual sources.

To the last point, I can only agree. A bookish mind, not surprisingly, seeks a bookish solution, though I hasten to add that the sort of trans-formation I envision entails affecting every sphere of our civilization, and therefore involves struggles on all fronts.

As for being ensconced in the educated elite, I can only plead guilty. The global environmental catastrophe is largely the making of this privileged stratum, whether in the form of science and technology or the brute fact that 70 percent of carbon dioxide emissions result from the top tenth of the world income earners (Milanovic 2016, 233).

In conclusion, I have no riposte save this book. The popular rendi-tion of chaos theory would have the flapping of butterfly wings, as they flutter by, leave profound effects in their wake. This metaphorical pebble hurled at a raging tsunami is unlikely to do much of anything, but in concert with other expressions and other actions, who knows?

APPENDIX

Modes of Denial and Procrastination

To the question—why are we so often inattentive and inactive even in the face of clear and present danger?—there are numerous answers (on the environmental crisis, see *inter alia* Hamilton 2010, Washington and Cook 2011, Marshall 2014, Ghosh 2016). Let me present here a conspectus of how we are prone to denial and procrastination.

Gazing at danger may trigger instinctive inaction, as in the proverbial deer in the headlight or the much maligned ostrich, falsely accused of sticking its head in the sand (cf. Rivers 1920, 55). There are probably evolutionary advantages to our reflexive immobility when we sense threat; a predator is more likely to spot a prey in motion, for instance. We know that a fire alarm in an auditorium or an airplane crash leads the majority to remain still and seated even if they realize the wisdom of a fight or flight response (cf. Cannon [1915] 1920, 197–203). Incredulity and immoblity are intimately intertwined. To be sure, conscious preparation can supersede instinctive responses, which are in turn manifold and often in tension with one another.

Beyond the biological level, mechanisms are in place to shroud clear and present danger. As an amplification of instinctive inaction, the psychology of denial trumps the reality principle. Indeed, psychoanalysis proposes denial as constitutive of the human psyche. We live without knowing what we know and may even wish not to know: to be unaware of our foundational impulses and desires, fundamental relations and conditions, and even irrefutable impressions and information. While repressing inconvenient truths, we carry on with convenient fictions. To paraphrase a twentieth-century poet, human beings cannot bear very much reality. We cannot overlook the ubiquity of psychological and sociological denial in everyday life (see Cohen 2001, Zerubavel 2006).

Our ethnocentric mindset focuses on a narrow circle of concerns, and the anticosmopolitan propensity underpins the hoary canard that all politics is local. The fire burning across the river is tantamount to the price of tea in China. In this regard, the father of economics Smith ([1759] 1976, 136) proposed a counterfactual of Chinese people being "swallowed up" by an earthquake. A moment of concern will wither away for an ordinary person but "if he was to lose his little finger tomorrow, he would not sleep to-night ... and the destruction of that immense multitude seems plainly an object less interesting to him, than this paltry misfortune of his own" (Smith [1759] 1976, 136–37; see Postface for his emendation). If it's not in my backyard, then it's none of my business. See no evil, hear no evil, speak no evil: these are said to be the attributes, after all, of a wise monkey.

If our ambit of care and concern is narrow, then our temporal orientation is out of joint. Many people are masterly in discounting long-term costs and benefits and exaggerating short-term rewards and punishments. Another sip of wine yields immediate, tangible pleasure, though most people are aware of the harm of repetitive and excessive drinking. We all know about addictive substances, but many disregard this information in thought and practice. Even for a committed environmentalist, the temptation to take a car for a long spin or to board another flight to a desirable destination makes for easy hypocrisy. The environmental crisis may be serious, but it seems just far enough off. As the mature Augustine ([397–400 CE] 2017, 223) recalled his youthful prayer: "Give me chastity and self-restraint, but don't do it just yet." *Sed noli modo*: Who cares about *mañana* when we are living *ahora*? This cognitive orientation may be hardwired or socially instilled, but its salience and prevalence seem incontrovertible.

More abstractly, we tend to regard change as minimal or gradual. To believe that the world is as it is—has been and will be—is a common outlook across diverse cultures and historical periods. If the world isn't quite stable, then we regard change as slow and long-lasting (perhaps forever). In so doing, we minimize the salience of multipliers and exponential growths or the possibility of cataclysmic transformations and apocalyptic ends (which does not contradict the persistent fascination many of us have with them). We have a cognitive propensity to presume family resemblance between the present and the future.

To the extent that we foresee change, we moderns believe in progress. From old-time religion to the culture of positivity, popular beliefs proffer bromides about the robustness of our existence and its pleasing prospects. We seem to have a bias for hope, and are irrationally positive (cf. Sharot 2011). The creed of positive psychology and the culture of bright(sided)ness privilege the happy-go-lucky in the merry-go-round of life that inevitably has its downs and its ups. Confronting bleak reality is depressing and debilitating, and without a measure of delusion, we may suffer from acedia or suicidal impulse. We may never achieve anything of substance if we don't undercalculate its costs. Happy science urges us to march on. Denial thus is functional when full accounting isn't imminent. Why fix something that isn't broken? Why worry about a catastrophe when we may have a eucatastrophe? We therefore wager on the irrefrangible nature of our habitat or expect a Hollywood-style deus ex machina to bring last-second rescue (only a few watch depressing art films, such as von Trier's *Melancholia* [2011], with their irrepressible propensity to prevent us from having a nice day). Positivity can be functional, but it can also be pathological. Cheerfulness may slide into irresponsibility, cool insouciance into reckless disregard, but it would take a heroic suspension of disbelief to expunge the indubitable threat and the inevitable tragedy. In general, conventional heuristics that facilitate everyday life decisions may generate grave outcomes in time of great danger, especially when they are unprecedented.

If the future is not yet, then the past is over and done with. We are resistant, and often immune, to moral tales of past disasters (Fressoz 2012). If it's shocking to read about how Europeans sleepwalked into World War I (Clark 2012), then it is all the more sobering to realize that Germans in the final year of World War II kept on fighting even when everyone believed that victory was impossible (Kershaw 1991). The lessons of history are frequently unheeded. We don't know much about history, and may therefore be doomed to repeat the same mistakes, and add new ones as well.

Political and economic forces not only wreak havoc on the environment, but also confound our grasp of reality. As Sinclair ([1935] 1994, 109) declared: "It is difficult to get a man to understand something, when his salary depends upon his not understanding it!" Owners and employees of coal mines or lobbyists who beseech politicians to pass legislation that promotes the antiquated, pollution-causing industry, for instance, not only endorse pro-coal ideology but also sponsor research that minimizes its deleterious consequences and sows confusion in public debate. The profit motive and the fossil ideology work hand in hand. Big science requires big money, and it should not be surprising that paymasters affect not only what people do research on, but also their findings and how they are disseminated. Here corporations bent on profit and scientists seeking new frontiers collude in perpetuating the problem (Oreskes and Conway 2010). The brute reality of influence-peddling and knowledge-distortion is something that every social scientist is aware of; the preponderance of evidence about global warming is immense and irrefutable, but the scientific consensus hardly

reaches the broad public. If only by sowing seeds of doubt, obnubila-
tion of the problem may occur. The cloud of unknowing hangs over
the world.

There is also a deeper assumption that animates professional and
amateur thinking on political economy: the primacy of growth. The
impeccable logic of growth—more for everyone—is a well-nigh unas-
sailable doctrine, but for a variety of reasons, including environmental
limitations, it is deeply flawed, however unpalatable the conclusion (Lie
2021, 40–57). Call it degrowth or nongrowth, along with a skeptical
take on the idea of progress, the rethinking of political economy is part
and parcel of a solution to the potential catastrophe that we face. Yet it
seems perverse and perhaps even nonsensical to most social scientists
(and the public at large), sustained by the twin dreams of progress and
growth.

Discounting the pain of others is a structural manifestation of indi-
vidual-level denial, which also constitutes the crux of economic and
policy analysis. From the standpoint of rational choice or instrumental
rationality, the propensity to seize immediate benefits is powerful: carpe
diem, as they say. If costs and consequences fall beyond our scope, then
why sacrifice ourselves for a good that is intangible and beyond the
horizon? In this regard, some argue that we possess the technologies
to stem global warming, but we have chosen not to do so (Pacala and
Socolow 2004). Why not enjoy the seemingly free ride of fossil-burning
transportation? The costs of environmental destruction are usually
borne by people living out of sight, including our own descendants,
and therefore remain out of our calculation, as externalities.

At the same time, we face a depressingly long list of existential
threats—from nuclear wars and volcanic eruptions to interstellar collisions and AI domination—that may preempt or precede the environmental threat (Ord 2020, 167). More concretely, right-thinking leftists
may worry that it would be indecent to expatiate on the not-yet crisis when there are concrete and urgent social problems. Why worry
about an intangible and invisible threat when there are injustices that
confront our immediate senses, such as abject poverty, and social and
spiritual problems, including alienation and anomie? Thus, it may not
be surprising that nature takes a back seat.

The matrix of denial may not only be congenital but also constructed
to ensure homophily of comforting beliefs and expunction of counter-
beliefs. While astutely avoiding deviant information, we enthusiasti-
cally embrace corroborating evidence . We believe what suits us, and
what suits us is what we believe. The logic of motivated reasoning—we
welcome confirmation bias and resist inconvenient refutations—may

be overdetermined by our social networks and the station to which we were thrust at birth (cf. Sloman and Fernbach 2017). Social media and other means of communication ensure that compatible and harmonious opinions surround and reverberate around us. A susurrus of caution is deafened by the loud and insistent gale of conventional wisdom. We are skilled at subsuming dissonant information and sustaining the received mindset. And our received belief systems—learned at home, in classroom, and through mass and social media—are remarkably robust, and they evinced little, if any, environmental concern until recently.

Everyday life, whether at home or work, is about the business of living, which is tantamount to busy-ness for most people. Given family obligations, career concerns, and other trials and tribulations of ordinary life, philosophical or metaphysical issues are often out of sight and therefore out of mind. The quiet desperation of ordinary getting and spending is naturalized and rendered as personal troubles that in turn are invisible and silent and don't transmogrify into public problems (Norgaard 2011, 60). Conservatism and complacency rule the roost of most people most of the time.

On a cynical note, we suffer from flaws—congential or cultivated—that have been identified since at least the beginning of literacy. We, both the fool and the wise, often enact stupidity: "The fool doth think he is wise, but the wise man knows himself to be a fool." As that most penetrating student on the topic wrote in an 1846 letter: "To be stupid, selfish, and in good health, those are the conditions required for happiness. But if you lack the first, all is lost" (Flaubert 1997, 63). Human stupidity is at once a cogent and disreputable explanation. It is not surprising, from this disenchanted perspective, that we are rakehells of our own doomed path to Ragnarök. Ignorance may be bliss, and few worry about the inevitable rain or storm when basking in sunshine. Not surprisingly, there are numerous studies of intelligence and other positive cognitive qualities, but agnotology remains an inchoate science (cf. Sternberg 2002, Proctor and Schiebinger 2008, Marmion 2019).

The happy few—perhaps the happy many—may be reluctant or incapable of facing up to the invisible threat. Paradoxically but predictably, we may suffer at once from impotence and arrogance. A complex problem that requires scientific literacy—reading, thinking, discussing, and reflecting—poses an enormous challenge that may reveal our intellectual impotence (cf. Jamieson 2014, 62–64). At the same time, powerlessness may transmogrify itself as arrogance, as ignorance and shortsightedness give people a mirage of omniscience, omnipotence, and therefore invulnerability. Whether we think fast or slow, we have a penchant for overconfidence.

These and other factors are significant, but they are not the focus of this book (though I touch on some of them). Certainly, books, articles, newsfeeds, podcasts, videos, and other sources of information have cascaded to the point where one may wonder if there's anything left to say (and an author cannot possibly acknowledge even a mere fraction of the vast scholarly output). Even Marx—the prophet of material progress and abundance, and therefore ostensibly a theorist inimical to environmental politics—has been resuscitated as an environmentalist *avant la lettre* (Foster 2000, Saito 2017). Even if everything *has* been said, however, we should probably keep on saying the same things, over and over again, if only to expiate our sins and to attempt to escape the trap of inattention and inaction.

REFERENCES

Abbott, Andrew. 2001. *Chaos of Disciplines*. Chicago: University of Chicago Press.

Ackerman, Jennifer. 2016. *The Genius of Birds*. New York: Penguin.

Adorno, Theodor W. (1957) 1975. "The Stars Down to Earth: The *Los Angeles Times* Astrology Column: A Study in Secondary Superstition." In Adorno, vol. 9.2 of *Gesammelte Schriften*, edited by Rolf Tiedemann. Frankfurt am Main: Suhrkamp.

Anders, Günther. 1956. *Die Antiquiertheit des Menschen: über die Seele im Zeitalter der zweiten industrielle Revolution*. Munich: Beck.

Aristotle. (c. 340 BCE) 2014. *Nicomachean Ethics*, rev. ed. Translated by Roger Crisp. Cambridge: Cambridge University Press.

Augustine. (397–400) 2017. *Confessions*. Translated by Sarah Ruden. New York: Modern Library.

Bacon, Lord. (1620) 1902. *Novum Organum*. Edited by Joseph Devey. New York: P. F. Collier & Son.

Ballard, J. G. 1964. *The Burning World*. New York: Berkley Books.

Bartlett, Robert. 2008. *The Natural and the Supernatural in the Middle Ages*. Cambridge: Cambridge University Press.

Beck, Ulrich. (1986) 1992. *Risk Society: Towards a New Modernity*. Translated by Mark Ritter. London: Sage.

Bell, J. S. 1964. "On the Einstein Podolsky Rosen Paradox." *Physics Physique Fizika* 1: 195–200.

Benedict, Barbara M. 2001. *Curiosity: A Cultural History*. Chicago: University of Chicago Press.

Benveniste, Émile. (1962) 2012. *Problèmes de linguistique générale*. Paris: Gallimard.

Bergson, Henri. (1889) 2001. *Time and Free Will: An Essay on the Immediate Data of Consciousness*. Translated by F. L. Pogson. Mineola, NY: Dover.

Bernard, Claude. (1927) 1957. *An Introduction to the Study of Experimental Medicine*. Translated by Henry Copley Greene. Mineola, NY: Dover.

Blastland, Michael, and David Spiegelhalter. 2013. *The Norm Chronicles: Stories and Numbers about Risk*. London: Profile Books.

Bloor, David. 2011. *The Enigma of the Aerofoil: Rival Theories in Aerodynamics, 1909–1930.* Chicago: University of Chicago Press.

Borchert, Donald M., ed. (1967) 2006. *Encyclopedia of Philosophy*, 2nd ed., 10 vols. Farmington Hills, MI: Thomson Gale.

Bourdieu, Pierre. 1972. *Esquisse d'une theorie de la pratique.* Paris: Droz.

Bourdieu, Pierre. 1979. *La Distinction: la critique sociale du jugement.* Paris: Éditions de Minuit.

Bowler, Peter J. 2001. *Reconciling Science and Religion: The Debate in Early-Twentieth-Century Britain.* Chicago: University of Chicago Press.

Boyer, Paul S., and Stephen Nissenbaum. 1974. *Salem Possessed: The Social Origins of Witchcraft.* Cambridge, MA: Harvard University Press.

Briggs, Robin. 1996. *Witches & Neighbors: The Social and Cultural Context of European Witchcraft.* New York: Viking.

Brzezinski, Zbigniew. 2013. "Afterword." In Jan Karski, *Story of a Secret State: My Report to the World.* Washington, DC: Georgetown University Press.

Broecker, Wallace S. 1975. "Climate Change: Are We on the Brink of a Pronounced Global Warming?" *Science* 189: 460–63.

Buell, Lawrence. 1995. *The Environmental Imagination: Thoreau, Nature Writing, and the Formation of American Culture.* Cambridge, MA: Harvard University Press.

Buharin N., and E. Preobrazhensky. (1920) 1922. *The ABC of Communism: A Popular Explanation of the Program of the Communist Party of Russia.* Translated by Eden Paul and Cedar Paul. London: The Communist Party of Great Britain.

Burnet, F. M. (1953) 1962. *Natural History of Infectious Diseases*, 2nd ed. Cambridge: Cambridge University Press.

Burnet, Macfarlane, and David O. White. (1953) 1972. *Natural History of Infectious Diseases*, 4th ed. Cambridge: Cambridge University Press.

Butler, Dom Cuthbert. (1922) 1926. *Western Mysticism: The Teaching of SS. Augustine, Gregory and Bernard on Contemplation and the Contemplative Life*, 2nd ed. London: E. P. Dutton.

Butler, Octavia E. 1993. *Parable of the Sower.* New York: Four Walls Eight Windows.

Butterfield, Herbert. 1950. *The Origins of Modern Science 1300–1800.* London: G. Bell and Sons.

Camic, Charles. 1986. "The Matter of Habit." *American Journal of Sociology* 91: 1039–87.

Camus, Albert. (1947) 2021. *The Plague.* Translated by Laura Morris. New York: Alfred A. Knopf.

Cannon, Walter B. [1915] 1920. *Bodily Changes in Pain, Hunger, Fear and Rage: An Account of Recent Researches into the Function of Emotional Excitement*, 2nd ed. New York: D. Appleton.

Carlisle, Clare. 2014. *On Habit.* London: Routledge.

Carpiano, Richard M. et al. 2023. "Confronting the Evolution and Expansion of Anti-Vaccine Activism in the USA in the Covid-19 Era." *The Lancet* 401: 967–70.

Carroll, Sean. 2022. *The Biggest Ideas in the Universe: Space, Time, and Motion*. New York: Dutton.

Carson, Rachel. 2018. *Silent Spring & Other Writings on the Environment*. Edited by Sandra Steingraber. New York: Library of America.

Cavell, Stanley. 1990. *Conditions Handsome and Unhandsome: The Constitution of Emersonian Perfectionism*. La Salle, IL: Open Court.

Chise, Diana, Margherita Fort, and Chiara Monfardini. 2020. "On the Intergenerational Transmission of STEM Education among Graduate Students." *B.E. Journal of Economic Analysis* 21: 115–45.

Cicchetti, Domenic V. 1991. "The Reliability of Peer Review for Manuscript and Grant Submissions: A Cross-Disciplinary Investigation." *Brain and Behavioral Sciences* 14: 119–35.

Clark, Christopher. 2012. *The Sleepwalkers: How Europe Went to War in 1914*. New York: Penguin.

Clark, Gregory, and David Jacks. 2007. "Coal and the Industrial Revolution, 1700–1869." *European Review of Economic History* 11: 39–72.

CMS.gov. 2021. "NHE Fact Sheet." https: //www.cms.gov/Research-Statistics-Data-and-Systems/Statistics-Trends-and-Reports/NationalHealthExpendData/NHE-Fact-Sheet.

Cohen, Joshua. 2013. *Attention! A (Short) History*. London: Nottingham Hill Editions.

Cohen, Stanley. 2001. *States of Denial: Knowing about Atrocities and Sufferings*. Cambridge: Polity Press.

Cohen, Stephen F. 1973. *Bukharin and the Bolshevik Revolution: A Political Biography, 1888–1938*. New York: Alfred A. Knopf.

Collingwood, R.G. 1924. *Speculum Mentis or the Map of Knowledge*. Oxford: Clarendon Press.

Collins, Harry, and Martin Kusch. 1998. *The Shape of Actions: What Humans and Machines Can Do*. Cambridge, MA: MIT Press.

Comte, Auguste. (1830) 1975. *Cours de philosophie positive*, 2 vols. Edited by Michel Serres, François Dagognet, Allal Sinaceur, and Jean-Paul Enthoven. Paris: Hermann.

Confucius. (c. 5th c. BCE) 2003. *Analects with Selections from Traditional Commentaries*. Translated by Edward Slingerland. Indianapolis, IN: Hackett.

Copenhaver, Brian. 2015. *Magic in Western Culture: From Antiquity to the Enlightenment*. Cambridge: Cambridge University Press.

Cowles, Henry M. 2020. *The Scientific Method: An Evolution of Thinking from Darwin to Dewey*. Cambridge, MA: Harvard University Press.

Crombie, A.C. (1952–59) 1961. *Augustine to Galileo*, 2nd ed. Cambridge, MA: Harvard University Press.

Crosby, Alfred W. 1976. *Epidemic and Peace: 1918*. Westport, CT: Greenwood Press.

Crosby, Alfred W. 1986. *Ecological Imperialism: The Biological Expansion of Europe, 900–1900*. Cambridge: Cambridge University Press.

Crutzen, Paul J. 2002. "Geology of Mankind." *Nature* 415: 23.

C-Span. 2002. "Defense Department Briefing," 12 February. https: // www.c-span.org/video/?168646–1/defense-department-briefing.

Danto, Arthur C. 1973. *Analytical Philosophy of Action*. Cambridge: Cambridge University Press.

Darnton, Robert. 1968. *Mesmerism and the End of the Enlightenment in France*. Cambridge, MA: Harvard University Press.

Darwin, Charles. (1859) 1869. *On the Origin of Species by Means of Natural Selection*, 5th ed. London: John Murray.

Daston, Lorraine. 1988. *Classical Probability in the Enlightenment*. Princeton, NJ: Princeton University Press.

Daston, Lorraine. 2019. *Against Nature*. Cambridge, MA: MIT Press.

De Pryck, Kari, and Mike Hulme, eds. 2023. *A Critical Assessment of the Intergovernmental Panel on Climate Change*. Cambridge: Cambridge University Press.

de Waal, Alex. 2021. *New Pandemics, Old Politics: Two Hundred Years of War on Disease and Its Alternatives*. Cambridge: Polity Press.

d'Entrèves, A. P. (1951) 1970. *Natural Law*, 2nd ed. London: Hutchinson.

Descola, Philippe. 2005. *Par-delà nature et culture*. Paris: Gallimard.

Donne, John. 1957. *The Sermons of John Donne*, vol. 3. Edited by George R. Potter and Evelyn M. Simpson. Berkeley: University of California Press.

Douglas, Mary, and Aaron Wildavsky. 1982. *Risk and Culture: An Essay on the Selection of Technological and Environmental Dangers*. Berkeley: University of California Press.

Duhem, Pierre. 1908. *SOZEIN TA PHAINOMENA, essai sur la notion de théorie physique de Platon à Galilée*. Paris: Hermann.

Edwards, Paul N. 2010. *A Vast Machine: Computer Models, Climate Data, and the Politics of Global Warming*. Cambridge, MA: MIT Press.

Einstein, Albert, Max Born, and Hedwig Born. 1971. *The Born–Einstein Letters: Correspondence between Albert Einstein and Max and Hedwig Born from 1916 to 1955*. Translated by Irene Born. New York: Walker.

Ellul, Jacques. 1954. *La technique ou l'enjeu du siècle*. Paris: Armand Colin.

Evans-Pritchard, E. E. 1937. *Witchcraft, Oracles and Magic among the Azande*. Oxford: Clarendon Press.

Falk, Seb. 2020. *The Light Ages: The Surprising Story of Medieval Science*. New York: W. W. Norton.

Ferguson, Cat, Adam Marcus, and Ivan Oransky. 2014. "Publishing: The Peer-Review Scam." *Nature* 515: 480–82.

Feyerabend, Paul. (1975) 1993. *Against Method*, 3rd ed. London: Verso.

Feynman, Richard. 1965. *The Character of Physical Law*. Cambridge, MA: MIT Press.

Fingarette, Herbert. (1972) 1998. *Confucius—The Secular as Sacred*. Long Grove, IL: Waveland Press.

Fingarette, Herbert. (1969) 2000. *Self-Deception*. Berkeley: University of California Press.

Flaubert, Gustave. 1997. *Selected Letters*. Translated by Geoffrey Wall. London: Penguin.

Flint, Valerie J. 1991. *The Rise of Magic in Early Medieval Europe*. Princeton, NJ: Princeton University Press.

Foster, John Bellamy. 2000. *Marx's Ecology: Materialism and Nature*. New York: Monthly Review Press.

Franklin-Wallis, Oliver. 2023. *Wasteland: The Dirty Truth about What We Throw Away, Where It Goes, and Why It Matters*. New York Simon & Schuster.

Frankopan, Peter. 2023. *The Earth Transformed: An Untold History*. New York: Alfred A. Knopf.

Freeman, Stephanie L. 2023. *Dreams for a Decade: International Nuclear Abolitionism and the End of the Cold War*. Philadelphia: University of Pennsylvania Press.

Fressoz, Jean-Baptiste. 2012. *L'apocalypse joyeuse: une histoire du risqué technologique*. Paris: Seuil.

Fry, Varian. 1942. "The Massacre of the Jews." *The New Republic*, 22 December, 816–19.

Frye, Northrop. 1964. *The Educated Imagination*. Bloomington: Indiana University Press.

Furue Takaharu. 2020. *Sugihara Chiune no jitsuzō: sūzennin no Yudayajin wo sukutta ketsudan to kakugo*. Tokyo: Mirutasu.

Galileo Galilei. (1632) 1890. "Dialogo sopra i due massimi." In *Galileo Galilei, Le Opere di Galileo Galilei*, vol. 7, edited by Antonio Favaro, Isidoro del Lungo, V. Cerrutti, G. Govi, G.V. Schiaparelli, and Umberto Marchesini. Florence: G. Barbèra.

Galison, Peter, and David J. Stump, eds. 1996. *The Disunity of Science: Boundaries, Contexts, and Power*. Stanford, CA: Stanford University Press.

Gallie, W. B. 1964. *Philosophy and the Historical Understanding*. London: Chatto & Windus.

Gamson, William A., Bruce Fireman, and Steven Rytina. 1982. *Encounters with Unjust Authority*. Homewood, IL: Dorsey.

Garrett, Laurie. 1994. *The Coming Plague: Newly Emerging Diseases in a World Out of Balance*. New York: Farrar, Straus & Giroux.

Geiger, Roger L. 2019. *American Higher Education since World War II: A History*. Princeton, NJ: Princeton University Press.

Ghosh, Amitav. 2016. *The Great Derangement: Climate Change and the Unthinkable*. Chicago: University of Chicago Press.

Giddens, Anthony. 1979. *Central Problems in Social Theory: Action, Structure and Contradiction in Social Analysis*. Berkeley: University of California Press.

Gigerenzer, Gerd, Zeno Swijtink, Theodore Porter, Lorraine Daston, John Beatty, and Lorenz Krüger. 1989. *The Empire of Chance: How Probability Changed Science and Everyday Life*. Cambridge: Cambridge University Press.

Gilson, Étienne. 1941. *God and Philosophy*. New Haven, CT: Yale University Press.

Gladstone, James, and Christy Lundy. 2018. *Earthrise: Apollo 8 and the Photo That Changed the World*. Berkeley: Owlkids Books.

Gödel, Kurt. 1931. "Über formal unentscheidbare Sätze der Principia Mathematica und verwandter Systeme, I." *Monatshefte für Mathematik und Physik* 38: 173–98.

Godfrey-Smith, Peter. 2020. *Metazoa: Animal Life and the Birth of the Mind*. New York: Farrar, Straus and Giroux.

Gollier, Christian. 2019. *Le climat après la fin du mois*. Paris: Presses Universitaires de France.

Graeber, David. 2018. *Bullshit Jobs: A Theory*. New York: Simon & Schuster.

Griffin, Donald R. (1992) 2001. *Animal Minds: Beyond Cognition and Consciousness*, rev. ed. Chicago: University of Chicago Press.

Hacking, Ian. 1990. *The Taming of Chance*. Cambridge: Cambridge University Press.

Hamilton, Clive. 2010. *Requiem for a Species: Why We Resist the Truth about Climate Change*. Abingdon: Earthscan.

Harford, Tim. 2020. *How to Make the World Add Up: Ten Rules for Thinking Differently about Numbers*. London: Bridge Street Press.

Harrison, Peter. 2015. *The Territories of Science and Religion*. Chicago: University of Chicago Press.

Helmholtz, Hermann von. (1891) 1912. "Autobiographical Sketches." In Helmholtz, *Popular Lectures on Scientific Subjects*, vol. 2. Translated by E. Atkinson. London: Longmans, Green.

Hobbes, Thomas. (1651) 1996. *Leviathan*. Edited by Richard Tuck. Cambridge: Cambridge University Press.

Hochschild, Arlie Russell. 1983. *The Managed Heart: Commercialization of Human Feeling*. Berkeley: University of California Press.

Horrobin, David F. 1990. "The Philosophical Basis of Peer Review and the Suppression of Innovation." *JAMA* 263: 1438–41.

Huber, Jürgen, Sabiou Inoua, Rudolf Kerschbamer, Chrisian König-Kersting, Stefan Palan, and Vernon L. Smith. 2022. "Nobel and Novice: Author Prominence Affects Peer Review." *PNAS* 119, no. 41. https://www.pnas.org/doi/10.1073/pnas.2205779119.

Hugh of St. Victor. (1176–77) 1961. *The Didascalicon of Hugh of St. Victor*. Translated by Jerome Taylor. New York: Columbia University Press.

Hume, David. (1739–40) 2007. *A Treatise on Human Nature*, vol. 1. Edited by David Fate Norton and Mary J. Norton. Oxford: Oxford University Press.

Huxley, Thomas Henry. 1893. *Darwiniana*. London: Macmillan.

Huxley, Thomas Henry. 1894. *Science and Christian Tradition: Essays*. New York: D. Appleton.

Ibn Rushd. (1178–79) 1961. "On the Harmony of Religion and Philosophy." Translated by George Hourani. https://www.aub.edu .lb/fas/CVSP/Documents/ReadingSelections/CVSP%20202/Fall %202012–2013/CVSP%20202%20Ibn%20Rushd.pdf.

Iliffe, Rob. 2017. *Priest of Nature: The Religious Worlds of Isaac Newton*. Oxford: Oxford University Press.

Institute for European Environmental Policy [IEEP]. 2020. "Green Deal for All." https://ieep.eu/uploads/articles/attachments/3b534d44 -4434-4ec7-af0b-7f6eb6c37882/Green%20Deal%20for%20All %20-%20FINAL%20PP.pdf?v=63756080686.

Intergovernmental Panel on Climate Change [IPCC]. 2001. "Third Assessment Report." https://www.ipcc.ch/assessment-report/ar3/.

Intergovernmental Panel on Climate Change [IPCC]. 2007. "Fourth Assessment Report." https: //www.ipcc.ch/assessment-report/ar4/.

Intergovernmental Panel on Climate Change [IPCC]. 2021. "Sixth Assessment Report." https: //www.ipcc.ch/report/ar6/wg1/.

Jacobsen, Anne M. 2015. *The Pentagon's Brain: An Uncensored History of DARPA, America's Top-Secret Military Research Agency*. New York: Back Bay Books.

Jamieson, Dale. 2014. *Reason in a Dark Time: Why the Struggle against Climate Change Failed—and What It Means for Our Future*. Oxford: Oxford University Press.

Jasanoff, Sheila. 1999. "The Songlines of Risk." *Environmental Values* 8: 135–52.

Jasanoff, Sheila. 2010. "A New Climate for Society." *Theory, Culture & Society* 27: 233–53.

Jaspers, Karl. 1958. *Die Atombombe und die Zukunft des Menschen: politisches Bewußtstein unserer Zeit*. Munich: R. Piper.

Jefferson, Tom, Philip Anderson, Elizabeth Wagner, and Frank Davidoff. 2002. "Effects of Editorial Peer Review: A Systematic Review." *JAMA* 287: 2784–86.

Jeffrey, Alice. 2023. "How Millenials and Gen Z Turned Astrology into a Billion-Dollar Industry." *Harper's BAZAAR*. https:// harpersbazaar.com.au/why-are-people-obsessed-with-astrology/.

Jencks, Christopher, and David Riesman. 1968. *The Academic Revolution*. Garden City, NY: Doubleday.

Jones, Eric M. 1985. *"Where Is Everybody?" An Account of Fermi's Question*. Los Alamos, NM: Los Alamos National Laboratory.

Jørgensen, Finn Arne. 2019. *Recycling*. Cambridge, MA: MIT Press.

Jungnickel, Christa, and Russell McCormmach. 1986. *The Intellectual Mastery of Nature: Theoretical Physics from Ohm to Einstein*, vol. 2. Chicago: University of Chicago Press.

Katz, Cheryl. 2019. "Piling Up: How China's Ban on Importing Waste Has Stalled Global Recycling." *Yale Environment 365*. https://e360.yale.edu/features/piling-up-how-chinas-ban-on-importing-waste-has-stalled-global-recycling.

Kaza, Stephanie, and Kenneth Kraft, eds. 2000. *Dharma Rain: Sources of Buddhist Environmentalism*. Boulder, CO: Shambhala.

Kelsen, Hans. 1946. *Society and Nature: A Sociological Inquiry*. London: Routledge.

Kennedy, Brian, and Meg Hefferon. 2019. *What Americans Know about Science*. Pew Research Center. https://pewrsr.ch/2V1RjGR.

Kershaw, Ian. 1991. *The End: The Defiance and Destruction of Hitler's Germany, 1944–45*. New York: Penguin.

Kline, Morris. 1953. *Mathematics in Western Culture*. Oxford: Oxford University Press.

Kline, Morris. 1980. *Mathematics: The Loss of Certainty*. New York: Oxford University Press.

Knight, Frank H. (1921) 1964. *Risk, Uncertainty and Profit*. New York: Augustus M. Kelley.

Kohn, Eduardo. 2013. *How Forests Think: Toward an Anthropology beyond the Human*. Berkeley: University of California Press.

Kokubun Kōichirō. 2017. *Chūdōtai no sekai: ishi to sekinin no kōkogaku*. Tokyo: Igaku Shoin.

Kolbert, Elizabeth. 2014. *The Sixth Extinction: An Unnatural History*. New York: Henry Holt.

Kolbert, Elizabeth. 2021. *Under a White Sky: The Nature of the Future*. New York': Crown.

Korsgaard, Christine M. 2018. *Fellow Creatures: Our Obligations to the Other Animals*. Oxford: Oxford University Press.

Kreiner, Jamie. 2023. *The Wandering Mind: What Medieval Monks Tell Us about Distraction*. New York: Liveright.

Kristeller, Paul Oskar. 1974. *Medieval Aspects of Renaissance Learning*. Edited by Edward P. Mahoney. Durham, NC: Duke University Press.

Kuhn, Thomas S. 1957. *The Copernican Revolution: Planetary Astronomy in the Development of Western Thought*. Cambridge, MA: Harvard University Press.

Kuhn, Thomas S. (1962) 2012. *The Structure of Scientific Revolutions*, 4th ed. Chicago: University of Chicago Press.

Lakatos, Imre. 1976. *Proofs and Refutations: The Logic of Mathematical Discovery*. Edited by John Worrall and Elie Zahar. Cambridge: Cambridge University Press.

Lakshmi, R.B. 2023. "The Environmental Impact of Battery Production for Electric Vehicles." https://earth.org/environmental-impact-of-battery-production/.

Lane, Melissa. 2011. *Eco-Republic: What the Ancients Can Teach Us about Ethics, Virtue, and Sustainable Living*. Witney: Peter Lang.

Lanzmann, Claude. 1978. "Karski Recalls His Meeting with United States Supreme Court Justice Felix Frankfurter, 1943." https: //art sandculture.google.com/asset/karski-recalls-his-meeting-with-unit

ed-states-supreme-court-justice-felix-frankfurter-1943-claude-lan
zmann/dgFS7OKiw9qznQ.

Laplace, Pierre Simon Marquis d'. (1814) 1902. *A Philosophical Essay on Probabilities*. Translated by Frederick Wilson Truscott and Frederick Lincoln Emory. London: John Wiley & Sons.

Latour, Bruno. 1991. *Nous n'avons jamais été modernes: Essai d'anthroplogie symétrique*. Paris: La Découverte.

Latour, Bruno. 2015. *Face à Gaia: Huit conferences sur le nouveau régime climatique*. Paris: La Découverte.

Laudan, Laurens. 1968. "Theories of Scientific Method from Plato to Mach: A Bibliographical Review." *History of Science* 7: 1–63.

Levi, Isaac. 1980. *The Enterprise of Knowledge: An Essay on Knowledge, Credal Probability, and Chance*. Cambridge, MA: MIT Press.

Lévi-Strauss, Claude. (1962) 1985. *Le totémisme aujourd'hui*, 6th ed. Paris: Presses Universitaires de France.

Lewis, Michael. 2003. *Moneyball: The Art of Winning an Unfair Game*. New York: W. W. Norton.

Lie, John. 2007. "Global Climate Change and the Politics of Disaster." *Sustainability Science* 2: 233–36.

Lie, John. 2021. *Japan, the Sustainable Society: The Artisanal Ethos, Ordinary Virtues, and Everyday Life in the Age of Limits*. Oakland: University of California Press.

Lindberg, David C. (1992) 2007. *The Beginnings of Western Science: The European Scientific Tradition in Philosophical, Religious, and Institutional Context, Prehistory to A.D. 1450*, 2nd ed. Chicago: University of Chicago Press.

Lotka, Alfred J. 1922. "Contribution to the Energetics of Evolution." *Proceedings of the National Academy of Sciences of the United States of America* 8: 147–51.

Louth, Andrew. 1981. *The Origins of the Christian Mystical Tradition: From Plato to Denys*. Oxford: Oxford University Press.

Lovelock, James. 1995. *The Ages of Gaia: A Biography of Our Living Earth*. New York: W. W. Norton.

Luhmann, Niklas. 1987. *Soziale Systeme: Grundriß einer allgemeinen Theorie*. Frankfurt am Main: Suhrkamp.

Lynas, Mark. 2020. *Our Final Warning: Six Degrees of Climate Emergency*. London: 4th Estate.

Malinowski, Bronislaw. (1948) 1954. *Magic, Science and Religion and Other Essays*. New York: Doubleday Anchor.

Malm, Andreas. 2016. *Fossil Capital: The Rise of Steam Power and the Roots of Global Warming*. London: Verso.

Mann, Janet. 2017. *Deep Thinkers: Inside the Minds of Whales, Dolphins, and Porpoises*. Chicago: University of Chicago Press.

Mann, Michael E. 2021. *The New Climate War: The Fight to Take Back Our Planet*. New York: PublicAffairs.

Mann, Michael E., and Tom Toles. 2016. *The Madhouse Effect: How Climate Change Denial Is Threatening Our Planet, Destroying Our*

Politics, and Driving Us Crazy. New York: Columbia University Press.

Mannheim, Karl. (1929) 1936. *Ideology ad Utopia: An Introduction to the Sociology of Knowledge.* Translated by Louis Wirth and Edward Shils. New York: Harcourt, Brace.

Manilii, M. (30–40 ce) 1903. *Astronomicon,* vol. 1. Edited by A. E. Housman. London: Grant Richards.

Manuel, Frank E. 1974. *The Religion of Isaac Newton.* Cambridge, MA: Harvard University Press.

Mark, Gloria. 2023. *Attention Span: A Groundbreaking Way to Restore Balance, Happiness and Productivity.* New York: Hanover Square Press.

Marmion, Jean-François, ed. 2019. *Histoire universelle de la connerie.* Paris: Sciences Humaines.

Marsh, George Perkins. 1864. *Man and Nature.* New York: Charles Scribner.

Marshall, George. 2014. *Don't Even Think About It: Why Our Brains Are Wired to Ignore Climate Change.* New York: Bloomsbury.

Marshall, T.H. 1950. *Citizenship and Social Class: And Other Essays.* Cambridge: Cambridge University Press.

Martin, Dale B. 2004. *Inventing Superstition: From the Hippocratics to the Christians.* Cambridge, MA: Harvard University Press.

Maudlin, Tim. 2011. *Quantum Non-Locality and Relativity: Metaphysical Intimacies of Modern Physics,* 3rd ed. Hoboken, NJ: Wiley-Blackwell.

Mayr, Ernst. 1982. *The Growth of Biological Thought: Diversity, Evolution, and Inheritance.* Cambridge, MA: Harvard University Press.

Mayr, Ernst. 1998. "Two Empires or Three?" *Proceedings of the National Academy of Sciences* 95: 9720–23.

Mazower, Mark. 2012. *Governing the World: The History of an Idea.* New York: Penguin.

McKibben, Bill. 1989. *The End of Nature.* New York: Random House.

McKibben, Bill, ed. 2008. *American Earth: Environmental Writing since Thoreau.* New York: Library of America.

Medoff, Rafael. 2022. *America and the Holocaust: Documentary History.* Philadelphia: Jewish Publication Society.

Mengzi. (c. 300 BCE) 2008. *Mengzi: With Selections from Traditional Commentaries.* Translated by Bryan W. Van Norden. Indianapolis, IN: Hackett.

Merton, Robert K. (1942) 1973. "The Normative Structure of Science." In Merton, *The Sociology of Science: Theoretical and Empirical Investigations.* Edited by Norman W. Storer. Chicago: University of Chicago Press.

Michaels, David. 2020. *The Triumph of Doubt: Dark Money and the Science of Deception.* New York: Oxford University Press.

Milanovic, Branko. 2016. *Global Inequality: A New Approach for the Age of Globalization*. Cambridge, MA: Harvard University Press.

Milgram, Stanley. 1971. *Obedience to Authority: The Experimental View*. New York: Harper & Row.

Mooney, Chris, and Sheril Kirschenbaum. 2009. *Unscientific American: How Scientific Illiteracy Threatens Our Freedom*. New York: Basic Books.

Moore, George Edward. 1903. *Principia Ethica*. Cambridge: Cambridge University Press.

Morton, Timothy. 2010. *The Ecological Thought*. Cambridge, MA: Harvard University Press.

Nadel, S. F. 1951. *The Foundations of Social Anthropology*. London: Cohen & West.

Needham, Joseph. 1956. *Science & Civilisation in China, Vol. 2: History of Scientific Thought*. Cambridge: Cambridge University Press.

Newman, John Henry, Cardinal. (1870) 1979. *An Essay in Aid of a Grammar of Assent*. Notre Dame, IN: University of Notre Dame Press.

Nhat Hanh, Thich. 1996. *The Miracle of Mindfulness: A Manual on Meditation*. Boston: Beacon Press.

NHK Merutodaun Shuzaihan. 2021. *Fukushima daiichi genpatsu jiko no "shinjitsu"*. Tokyo: Kōdansha.

Nietzsche, Friedrich. (1886) 1998. *Beyond Good and Evil*. Translated and edited by Marion Faber. Oxford: Oxford University Press.

Noble, David F. 1997. *The Religion of Technology: The Divinity of Man and the Spirit of Invention*. New York: Alfred A. Knopf.

Nock, Arthur Darby. 1933. *Conversion: The Old and the New in Religion from Alexander the Great to Augustine of Hippo*. London: Oxford University Press.

Nock, Arthur Darby. 1972. *Essays on Religion and the Ancient World*, vol. 2. Edited by Zeph Stewart. Cambridge, MA: Harvard University Press.

Nordhaus, William D. 2021. *The Spirit of Green: The Economics of Collisions and Contagions in a Crowded World*. Princeton, NJ: Princeton University Press.

Norgaard, Kari Marie. 2011. *Living in Denial: Climate Change, Emotions, and Everyday Life*. Cambridge, MA: MIT Press.

North, John. (1993) 2008. *Cosmos: An Illustrated History of Astronomy and Cosmology*. Chicago: University of Chicago Press.

Northcott, Robert, and Anna Alexandrova. 2015. "Prisoner's Dilemma Doesn't Explain Much." In *The Prisoner's Dilemma*, edited by Martin Peterson. Cambridge: Cambridge University Press.

Nuclear Threat Initiative. 2019. *Global Health Security Index*. https://www.ghsindex.org/wp-content/uploads/2021/11/2019-Global-Health-Security-Index.pdf.

Nuland, Sherwin B. 2003. *The Doctors' Plague: Germs, Childhood Fever, and the Strange Story of Ignác Semmelweis.* New York: W. W. Norton.

Nussbaum, Martha. 2023. *Justice for Animals: Our Collective Responsibility.* New York: Simon & Schuster.

Nye, David E. 1994. *American Technological Sublime.* Cambridge, MA: MIT Press.

Offill, Jenny. 2020. *Weather.* New York: Alfred A. Knopf.

Ord, Toby. 2020. *The Precipice: Existential Risk and the Future of Humanity.* New York: Hachette Books.

Oreskes, Naomi. 2019. *Why Trust Science?* Princeton, NJ: Princeton University Press.

Oreskes, Naomi. 2021. *Science on a Mission: How Military Funding Shaped What We Do and Don't Know about the Ocean.* Chicago: University of Chicago Press.

Oreskes, Naomi, and Erik M. Conway. 2010. *Merchants of Doubt: How a Handful of Scientists Obscured the Truth on Issues from Tobacco Smoke to Global Warming.* New York: Bloomsbury Press.

Oreskes, Naomi, and Erik M. Conway. 2014. *The Collapse of Western Civilization: A View from the Future.* New York: Columbia University Press.

Ortega y Gasset, José. (1930) 1932. *The Revolt of the Masses.* Translation anonymous. New York: W. W. Norton.

Osborn, Fairfield. 1948. *Our Plundered Planet.* Boston: Little, Brown.

Ostrogorski, M. 1902. *Democracy and the Organization of Political Parties,* vol. 2. Translated by Frederick Clarke. London: Macmillan.

Pacala, S., and R. Socolow. 2004. "Stabilization Wedges: Solving the Climate Problem for the Next 50 Years with Current Technologies." *Science* 305: 968–72.

Page, Don N., and William K. Wootters. 1983. "Evolution without Evolution: Dynamics Demonstrated by Stational Observables." *Physical Review D* 27: 2885–92.

Parker, Geoffrey. 2013. *Global Crisis: War, Climate Change & Catastrophe in the Seventeenth Century.* New Haven, CT: Yale University Press.

Parsons, Talcott. 1951. *The Social System.* Glencoe, IL: Free Press.

Pascal, Blaise. (1670) 1995. *Pensées.* Translated by Honor Levi. Oxford: Oxford University Press.

Passmore, John. 1974. *Man's Responsibility for Nature: Ecological Problems and Western Traditions.* New York: Charles Scribner's Sons.

Pater, Walter. [1873] 1888. *The Renaissance: Studies in Art and Poetry.* London: Macmillan.

Pindyck, Robert S. 2022. *Climate Future: Averting and Adapting to Climate Change.* Oxford: Oxford University Press.

Plain English Campaign. 2003. "Foot in Mouth Award 2003." http://www.plainenglish.co.uk/campaigning/awards/2001–2010-awards/2003-awards/811-foot-in-mouth-award-2003.html.

Planck, Max. 1948. *Die wissenschaftlicher Selbstbiographie.* Leipzig: Johann Ambrosius Barth.

Polkinghorne, John. 2007. *Quantum Physics and Theology: An Unexpected Kinship.* New Haven, CT: Yale University Press.

Pomeranz, Kenneth. 2000. *The Great Divergence: China, Europe, and the Making of the Modern World Economy.* Princeton, NJ: Princeton University Press.

Porter, Theodore M. 1986. *The Rise of Statistical Thinking 1820–1900.* Princeton, NJ: Princeton University Press.

Proctor, Robert. 2012. *Golden Holocaust: Origins of the Cigarette Catastrophe and the Case for Abolition.* Berkeley: University of California Press.

Proctor, Robert N., and Londa Schiebinger, eds. 2008. *Agnotology: The Making and Unmaking of Ignorance.* Stanford, CA: Stanford University Press.

Proust, Marcel. (1925) 2002. *The Fugitive.* Translated by Peter Collier. London: Allen Lane.

Quammen, David. 2018. *The Tangled Tree: A Radical New History of Life.* New York: Simon & Schuster.

Quine, W. V., and J. S. Ullian. 1970. *The Web of Belief.* New York: Random House.

Ravaisson, Félix. 1838. *De l'habitude.* Paris: H. Fournier.

Revelle, Roger, and Hans E. Suess. 1957. "Carbon Dioxide Exchange between Atmosphere and Ocean and the Question of an Increase of Atmospheric CO2 during the Past Decades." *Tellus* 9: 18–27.

Rich, Nathaniel. 2019. *Losing Earth: A Recent History.* New York: Farrar, Straus and Giroux.

Ridley, Matt, and Alina Chan. 2021. *Viral: The Search for the Origin of COVID-19.* New York: Harper.

Rigden, John S. 2005. *Einstein 1905: The Standard of Greatness.* Cambridge, MA: Harvard University Press.

Ritchie, Stuart. 2020. *Science Fictions: How Fraud, Bias, Negligence, and Hype Undermine the Search for Truth.* New York: Metropolitan Books.

Rivers, W.H.R. 1920. *Instinct and the Unconscious: A Contribution to a Biological Theory of the Psycho-neuroses.* Cambridge: Cambridge University Press.

Robinson, Kim Stanley. 2020. *The Ministry for the Future.* New York: Orbit.

Royce, Josiah. 1913. *The Problem of Christianity*, vol. 2. New York: Macmillan.

Russell, Bertrand. 1951. "Ludwig Wittgenstein." *Mind* 60: 297–98.

Ryle, Gilbert. 1945–46. "Knowing How and Knowing That: The Presidential Address." *Proceedings of the Aristotelian Society* 46: 1–16.

Sabel, Charles F., and David G. Victor. 2022. *Fixing the Climate: Strategies for an Uncertain World.* Princeton, NJ: Princeton University Press.

Sahlins, Marshall. 2008. *The Western Illusion of Human Nature.* Chicago: Prickly Paradigm Press.

Saito, Kohei. 2017. *Karl Marx's Ecosocialism: Capital, Nature, and the Unfinished Critique of Political Economy.* New York: Monthly Review Press.

Sambursky, Samuel. (1956) 1987. *The Physical World of the Greeks,* 2nd ed. Translated by Merton Dagut. Princeton, NJ: Princeton University Press.

Sapp, Jan A. 2009. *The New Foundations of Evolution: On the Tree of Life.* New York: Oxford University Press.

Sarna, Nahum P. 1989. *Genesis: The Traditional Hebrew Text with New JPS Translation / Commentary.* Philadelphia: The Jewish Publication Society.

Schell, Jonathan. 2020. *The Fate of the Earth, The Abolition, The Unconquerable World.* Edited by Martin J. Sherwin. New York: Library of America.

Schopenhauer, Arthur. (1841) 2017. "Preisschrift über die Grundlage der Moral." In Schopenhauer, *Die beiden Grundprobleme der Ethik.* Edited by Angelika Hübscher. Zurich: Diogenes.

Schweber, Libby. 2006. *Disciplining Statistics: Demography and Vital Statistics in England and France, 1830–1885.* Durham, NC: Duke University Press.

Schweber, Silvan S. 2000. *In the Shadow of the Bomb: Oppenheimer, Bethe, and the Moral Responsibility of the Scientist.* Princeton, NJ: Princeton University Press.

Servigne, Pablo, and Raphaël Steven. 2015. *Comment tout peut s'effordrer: petit manuel de collapsologie à l'usage des générations presents.* Paris: Seuil.

Shapin, Steven. 1993. *A Social History of Truth: Civility and Science in Seventeenth-Century England.* Chicago: University of Chicago Press.

Shapin, Steven. 1996. *The Scientific Revolution.* Chicago: University of Chicago Press.

Shapin, Steven, and Simon Schaffer. 1985. *Leviathan and the Air-Pump: Hobbes, Boyle, and the Experimental Life.* Princeton, NJ: Princeton University Press.

Sharot, Tali. 2011. *The Optimism Bias: A Tour of the Irrationally Positive Brain.* New York: Pantheon.

Sharpe, Simon. 2023. *Five Times Faster: Rethinking the Science, Economics, and Diplomacy of Climate Change.* Cambridge: Cambridge University Press.

Shatz, David. 2004. *Peer Review: A Critical Inquiry.* Lanham, MD: Rowman & Littlefield.

Sherwin, Martin J. 2020. *Gambling with Armageddon: Nuclear Roulette from Hiroshima to the Cuban Missile Crisis.* New York: Alfred A. Knopf.

Shklar, Judith N. 1990. *The Faces of Injustice.* New Haven, CT: Yale University Press.

Shklovsky, Victor. [1917] 1965. "Art as Technique." In *Russian Formalist Criticism: Four Essays*. Translated by Lee T. Lemon and Marion J. Reis. Lincoln: University of Nebraska Press.

Sinclair, Upton. (1935) 1994. *I, Candidate for Governor: And How I Got Licked*. Berkeley: University of California Press.

Sloman, Steven, and Philip Fernbach. 2017. *The Knowledge Illusion: Why We Never Think Alone*. New York: Riverhead.

Smil, Vaclav. 1993. *Global Ecology: Environmental Change and Social Flexibility*. London: Routledge.

Smil, Vaclav. 2017. *Energy and Civilization: A History*. Cambridge, MA: MIT Press.

Smil, Vaclav. 2022. *How the World Really Works*. New York: Viking.

Smith, Adam. (1759) 1976. *The Theory of Moral Sentiments*. Edited by D. D. Raphael and A. L. Macfie. Oxford: Clarendon Press.

Smith, Richard. 2006. "Peer Review: A Flawed Process at the Heart of Science and Journals." *Journal of the Royal Society of Medicine* 99: 178–82.

Smith, Wilfred Cantwell. 1963. *The Meaning and End of Religion: A New Approach to the Religious Traditions of Mankind*. New York: Macmillan.

Snowden, Frank M. 2019. *Epidemics and Society: From the Black Death to the Present*. New Haven, CT: Yale University Press.

Sombart, Werner. 1902. *Der moderne Kapitalismus*, 2 vols. Leipzig: Duncker & Humblot.

Sonenscher, Michael. 2022. *Capitalism: The Story Behind a Word*. Princeton, NJ: Princeton University Press.

Spier, Ray. 2002. "The History of the Peer-Review Process." *Trends in Biotechnology* 20: 357–58.

Stanier, R. Y., and C. B. van Niel. 1962. "The Concept of a Bacterium." *Arkiv für Mikrobiologie* 42: 17–35.

Stedman Jones, Gareth. 2004. *An End to Poverty? A Historical Debate*. New York: Columbia University Press.

Stepanova, Maria. 2021. *In Memory of Memory*. Translated by Sasha Dugdale. New York: New Directions.

Stern, Nicholas. 2006. *The Economics of Climate Change: The Stern Review*. Cambridge: Cambridge University Press.

Stern, Nicholas. 2015. *Why Are We Waiting? The Logic, Urgency, and Promise of Tackling Climate Change*. Cambridge, MA: MIT Press.

Sternberg, Robert J., ed. 2002. *Why Smart People Can Be So Stupid*. New Haven, CT: Yale University Press.

Strauss, Leo. 1970. *Xenophon's Socratic Discourse: An Interpretation of the Oeconomicus*. Ithaca, NY: Cornell University Press.

Stroebe, Wolfgang, Tom Postmes, and Russell Spears. 2012. "Scientific Misconduct and the Myth of Self-Correction in Science." *Perspectives on Psychological Science* 7: 670–88.

Tainter, Joseph A. 1988. *The Collapse of Complex Societies*. Cambridge: Cambridge University Press.

Tarski, Alfred. (1956) 1983. *Logic, Semantics, Metamathematics: Papers from 1923 to 1938*, 2nd ed. Edited by John Corcoran and translated by J. E. Woodger. Indianapolis, IN: Hackett.

Tec, Nechama. 2013. *Resistance: Jews and Christians Who Defied the Nazi Terror*. Oxford: Oxford University Press.

Tetlock, Philip E. 2005. *Expert Political Judgment: How Good Is It? How Can We Know?* Princeton, NJ: Princeton University Press.

Thaler, Richard H. 2016. "Behavioral Economics: Past, Present, and Future." *American Economic Review* 106: 1567–1600.

Thomas Aquinas, Saint. (1470) 1933. *Scriptum Super Sententiis*. Paris: Lethielleux.

Thomas Aquinas, Saint. (1485) 2012. *Summa Theologiae Prima Secundae, 1–70*. Edited by the Aquinas Institute and translated by Fr. Laurence Shapcote. Green Bay, WI: Aquinas Institute.

Thomas, Keith. (1971) 1973. *Religion and the Decline of Magic*. London: Penguin.

Thomas, Keith. 1983. *Man and the Natural World: A History of the Modern Sensibility*. New York: Pantheon Books.

Thompson, E.P. 1980. "Notes on Exterminism, the Last Stage of Civilisation." *New Left Review* 121: 3-31.

Trevor-Roper, H. R. 1959. "The General Crisis of the Seventeenth Century." *Past & Present* 16: 31–64.

Trotsky, Leon. (1930) 2017. *The History of the Russian Revolution*, vol.1. Translated by Max Eastman. https: //www.marxists.org/archive/trotsky/works/download/hrr-vol1.pdf.

United Health Foundation. 2022. "America's Health Rankings 2021." https: //www.americashealthrankings.org/explore/annual/measure/PH_funding/state/ALL.

Vaan, Michiel de. 2008. *Etymological Dictionary Latin and the Other Italic Languages*. Leiden: Brill.

Vaihinger, Hans. 1911. *Die Philosophie des als ob: System der theoretischen, praktischen und religiösen Fiktionen der Menschheit auf Grund eines idealitischen Positivismus*. Leipzig: Felix Meiner.

van Fraassen, Bas C. 1989. *Laws and Symmetry*. Oxford: Clarendon Press.

Veblen, Thorstein. 1918. *The Higher Learning in America: A Memorandum on the Conduct of Universities by Business Men*. New York: B. W. Huebsch.

Vogt, William. 1948. *Road to Survival*. New York: William Sloane Associates.

Vollmann, William T. 2018. *Carbon Ideologies*, vol. 1. New York: Viking.

Walker, D. P. 1958. *Spiritual and Demonic Magic from Ficino to Campanella*. London: Warburg Institute.

Walkley, Sarah. 2023. "The Carbon Cost of Social Media." https://carbonliteracy.com/the-carbon-cost-of-social-media/#:~:text=On%20average%2C%20we%20spend%20145,Edinburgh%20in%20a%20small%20car.

Wallace-Wells, David. 2019. *The Uninhabitable Earth: Life After Warming.* New York: Tim Duggan Books.

Wallace-Wells, David. 2021. "How to Live in a Climate 'Permanent Emergency.'" *New York*, July. https: //nymag.com/intelligencer /2021/07/how-to-live-in-a-climate-permanent-emergency.html.

Wallerstein, Immanuel, et al. 1996. *Open the Social Sciences: Report of the Gulbenkian Commission on the Restructuring of the Social Sciences.* Stanford, CA: Stanford University Press.

Wallerstein, Immanuel. 2006. *European Universalism: The Rhetoric of Power.* New York: New Press.

Washington, Haydn, and John Cook. 2011. *Climate Change Denial: Heads in the Sand.* Abingdon.: Earthscan.

Weart, Spencer R. 2008. *The Discovery of Global Warming.* Cambridge, MA: Harvard University Press.

Webb, Stephen. (2002) 2015. *If the Universe is Teeming with Aliens ... WHERE IS EVERYBODY? Seventy-Five Solutions to the Fermi Paradox and the Problem of Extraterrestrial Life,* 2nd ed. Cham, SW: Springer International Publishing.

Weber, Max. (1904–5) 1930. *The Protestant Ethic and the Spirit of Capitalism.* Translated by Talcott Parsons. London: George Allen & Unwin.

Weber, Max. (1917) 2004. *The Vocation Lectures.* Edited by David Owen and Tracy B. Strong, and translated by Rodney Livingstone. Indianapolis, IN: Hackett.

Weber, Max. (1921–22) 2019. *Economy and Society: A New Translation.* Edited and translated by Keith Tribe. Cambridge, MA: Harvard University Press.

Wells, H. G. 1895. *The Time Machine.* London: Heinemann.

Wells, H. G. 1920. *The Outline of History: Being a Plain History of Life and Mankind,* vol. 2. London: Macmillan.

Welzer, Harald. 2008. *Klimakriege: Wofür im 21. Jahrhundert getötet wird.* Frankfurt am Main: S. Fischer.

Whewell, William. 1860. *On the Philosophy of Discovery.* London: John Parker & Son.

Whitehead, Alfred North. 1920. *The Concept of Nature.* Cambridge: Cambridge University Press.

Whitehead, Alfred North. 1929. *Science and the Modern World.* Cambridge: Cambridge University Press.

Whitehead, Alfred North, and Bertrand Russell. 1912. *Principia Mathematica,* vol. 2. Cambridge: Cambridge University Press.

Winner, Langdon. (1986) 2020. *The Whale and the Reactor: A Search for Limits in an Age of High Technology,* 2nd ed. Chicago: University of Chicago Press.

Witten, Edward. 1995. "String Theory Dynamics in Various Dimensions." *Nuclear Physics B* 443: 85–126.

Wittgenstein, Ludwig. (1956) 1978. *Remarks on the Foundation of Mathematics,* rev. ed. Edited by G. H. von Wright, R. Rhees, and G. E. M. Anscombe, and translated by G. E. M. Anscombe. Cambridge, MA: MIT Press.

Wittgenstein, Ludwig. 1969. *On Certainty*. Edited by G. E. M. Anscombe and G. E. M. H. von Wright, and translated by Denis Paul and G. E. M. Anscombe. Oxford: Blackwell.

Woese, Carl R., and George E. Fox. 1977. "Phylogenetic Structure of the Prokaryotic Domain: The Primary Kingdoms." *Proceedings of the National Academy of Sciences* 74: 5088–90.

Woese, Carl R., Otto Kandler, and Mark L. Wheelis. 1990. "Toward a Natural System of Organisms: Proposal for the Domains Archaea, Bacteria, and Eucarya." *Proceedings of the National Academy of Sciences* 87: 4576–79.

Wohlleben, Peter. 2015. *Das geheime Leben der Bäume: Was sie fühlen, wie sie kommunizieren*. Munich: Ludwig Verlag.

Wyman, David S. 1984. *The Abandonment of the Jews: America and the Holocaust, 1941–1945*. New York: Pantheon.

Yamamoto Yoshitaka. 2015. *Genshi, genshikaku, genshiryoku: watashi ga kōgi de tsutaetakatta koto*. Tokyo: Iwanami Shoten.

Yamauchi Mikio. 2012. "Fukushima to sono shūhen ni nokoru rekishi tsunami no kiroku." http: //yamagatuko.sakura.ne.jp/tunami.html.

Yates, Frances A. 1964. *Giordano Bruno and the Hermetic Tradition*. London: Routledge & Kegan Paul.

Yates, Frances A. 1972. *The Rosicrucian Enlightenment*. London: Routledge & Kegan Paul.

Young, Michael W. 2004. *Malinowski: Odyssey of an Anthropologist, 1884–1920*. New Haven, CT: Yale University Press.

Yurchak, Alexei. 2005. *Everything Was Forever, Until It Was No More: The Last Soviet Generation*. Princeton, NJ: Princeton University Press.

Zagorin, Perez. 1998. *Francis Bacon*. Princeton, NJ: Princeton University Press.

Zerubavel, Eviatar. 2006. *The Elephant in the Room: Silence and Denial in Everyday Life*. New York: Oxford University Press.

Zuckerman, Gregory. 2019. *The Man Who Solved the Market: How Jim Simons Launched the Quant Revolution*. New York: Portfolio.

Zweig, Stefan. (1927) 1964. *Sternstunden der Menschheit: Vierzehn historische Miniaturen*. Frankfurt am Main: S. Fischer.

INDEX

abstraction: Bergson on 61; and change 62–63; conceptual 14; division 8; irrelevant 47; in knowledge 13, 47, 53, 63; in theory 34

action: -centric theory 54, 56–58; conscious xii–xiii, 53, 58–59; consequence of 41; emotional 54; instrumentally rational 54; knowledge and xiv, 47–48, 52–53, 60, 62, 72; lag in 13, 41, 45; meaningful 54, 56; rare 55; rational xiii, 54; reflective 47; saving the phenomenon of 56; social 56; source of 53; strategic 57; and structure 53–58; substantively rational 54; therapeutic 62; traditional 54; and transformation 60, 73; Weber's typology of 54–55; *see also* inaction

Acts, Book of 51

Adorno, Theodor W.: quoted 31–32

Age of Reason 2, 9, 35

agency *see* action

agnotology *see* stupidity

airplane 33, 36

American Economic Association 11

animal: human affinity with xi, 25; in general 35; mind 43; rights 44; savagery toward 43–44; species 43

Annalen der Physik (journal) 16

Anthropocene 8

anthropocentrism: in general x; "man the measure" 13

anthropology 8–9

Apollo 8, 66

Aquinas *see* St. Thomas Aquinas

Arabic: natural philosophy 31

Archaea 13–15, 32

Arianism 36

Aristotle: on *hexis* 60–61; his natural philosophy 5, 31; quoted 60

Arrhenius, Svante 1, 10

arrogance *see* hubris

Art 65

as if *see* counterfactual

Asia 7

Association for Environmental Studies and Sciences 11

astrology 30–32, 34

astronomy 32, 34

Attention Deficit Hyperactivity Disorder (ADHD) 55

authority: academic 5, 8; epistemic 18, 35; in general 15; of knowledge 15, 27; religious 15; scientific 12; of the scientific method 4; status-based 12, 27, 68, 70

automobile *see* car

Azande (Sudan) 29–33